The Land of the Beacons Way:
Scenery and geology across the Brecon Beacons National Park

For everyone interested in the evolution of this landscape.

By Dilw...

© Geologists' Association - South Wales Group
Cymdeithas y Daearegwyr - Grŵp De Cymru

www.swga.org.uk

2014

ISBN 978-0-903222-02-0
Geologists' Association - South Wales Group,
Department of Geology,
Amgueddfa Cymru - National Museum Wales, Cardiff
CF10 3NP

Cover illustration:
View towards Pen y Fan from Craig Cerrig-gleisiad. Photo by Andy Kendall

Contents

Preface	5
The landscape	5
A brief introduction to geology	6
Part 1. The big picture	11

Part 2. The walks

Day 1	Ysgyryd Fawr (The Skirrid) to Llanthony	34
Day 2	Llanthony to Crickhowell	44
Day 3	Crickhowell to Llangynidr	51
Day 4	Llangynidr to Storey Arms	58
Day 5	Storey Arms to Craig-y-nos	67
Day 6	Craig-y-nos to Llanddeusant	80
Day 7	Llanddeusant to Carreg Cennen	88
Day 8	Carreg Cennen to Bethlehem	103

Glossary	111
Further Information	119
Acknowledgements	120
Map Key	128

Safety advice
This volume is not a guide to the route which is described in detail in The Beacons Way by John Sanson and Arwel Michael, 2nd edition 2011. Most days are strenuous walks over exposed country. Good navigation skills, the ability to read a map and to use a compass are essential as there is no waymarking on open hills and moorland, only over the lower areas, and some of the paths are not obvious (especially in poor visibility). Good walking maps are essential, eg. OS 1:25,000 Explorer maps OL 12 and 13, or equivalent.

The route was devised and developed by two members of the Brecon Beacons Park Society and there is additional information on the society's website, www.breconbeaconsparksociety.org. Follow the link under Walks to Treks and Trails/Beacons Way and include the Hints & Tips link under: The Route, which gives essential advice and details of route changes and errors on the Ordnance Survey maps.

You must be properly equipped and aware of the fact that the weather can change rapidly for the worse. A good source of weather information is the Met Office mountain area forecast for the Brecon Beacons.

It is rarely necessary to leave the path to see the features described. If you do leave the route take great care on steep slopes and rough, loose or slippery rocky surfaces. Avoid the edges of steep drops and anywhere where rocks could fall on youor others. Do not climb rock faces or cliffs. You are responsible for your own safety.

Please take great care not to damage rock faces, walls, buildings or ancient monuments or to trample vegetation. If you must collect rock specimens please only do so from loose material, but it is better just to take photographs.

Geological maps
A geological map is not essential but can help in understanding the features. The relevant maps for each day are listed. Do not rely on them for navigation.

Part of Day 4 and all of Days 5 – 8 are also covered by a 1:50,000 map published in 2014 by British Geological Survey for Fforest Fawr Geopark (*Fforest Fawr, Exploring the Landscape of a Global Geopark*). This map is for the non-specialist and provides an overview and interpretation of the bedrock and superficial deposits plus information about selected sites.

Geological terms and especially geological formation names are frequently updated. The terms used in this book are those that are most likely to be found in other maps and books that you will find describing the geology of this area.

Preface

The Beacons Way is a 152 kilometre (95 mile), linear, long distance walking route crossing the Brecon Beacons National Park from Ysgyryd Fawr in the east to Bethlehem in the west. Walkers usually divide the Way into 8 stages, which can obviously be walked singly or consecutively.

As the Beacons Way traverses the entire Brecon Beacons National Park, this book is also a description of most of the geological and landscape history of the National Park.

The western section of the Way passes through Fforest Fawr Geopark, a member of the European Geoparks and Global Geoparks Networks. This area is special not only because of its outstanding geological history, rocks and landscape but also on account of the archaeology, industrial heritage, wildlife and culture. There is much more information on the Fforest Fawr Geopark website: www.fforestfawrgeopark.org.uk

Prior geological knowledge is not essential, but always useful.

Most of the geological terms used are explained in the glossary on page 111.

The landscape

The land you tread along the Beacons Way, from Ysgyryd Fawr to Bethlehem, has been about 470 million years in the making. In that time South Wales has journeyed more than 12,000 km, from 70° south to 50° north. On the way it has twice been caught up in the mill of colliding continents and then spat out again as these continents ripped apart. Beginning our story as a muddy sea it has become in turn: coastal semi-arid desert plain, clear sun-lit tropical sea, river delta, tropical swamp, desert again, island archipelago, possibly a deep sea floor thick with chalk, and then was ice covered for almost 2.6 million years. Since the ice melted South Wales has continued to change, possibly faster than ever, as human occupation and use supervene on slow but massive geological processes.

Part 1 is a brief geological history of this area, the big picture of what was happening both globally and locally, and why. It is a simplified version of a very complex story, some of which is not yet fully understood, and only covers what is relevant to the Beacons Way. All dates are reasonable approximations. When a country is mentioned, eg South Wales, the reference is to the area that will eventually become South Wales.

Part 2 describes the easily visible and interesting features on each day's walk, given decent visibility.

An introduction to geology is included for those who might appreciate it.

A brief introduction to geology

What are rocks?

A rock is a naturally occurring solid assemblage of grains. These grains can be fragments of pre-existing rocks, shell fragments or mineral crystals. A mineral is also a naturally occurring solid substance but it is crystalline and has a fixed atomic structure. The building blocks of minerals are elements, such as silica, oxygen, calcium, aluminium, iron etc.

Sandstones are predominantly made of sand grains derived from older rocks. Sand grains themselves are composed of mineral crystals (mainly quartz and feldspar)

Mudstones are assemblages of fine grained clay mineral crystals.

Limestones are collections of calcium carbonate grains, either as shell fragments or as crystals of various sizes.

Types of rocks

Three types of rocks are found on the Earth's surface, igneous, metamorphic and sedimentary. Igneous rocks, such as granite and basalt, crystallize from molten magma or lava as it cools. Metamorphic rocks are any rock that has been changed (re-crystallized) from its original form by heat and/or pressure. Slate is a metamorphic rock which started life as a mudstone. Marble is another metamorphic rock, formed from limestone. Sedimentary rocks, such as sandstone, mudstone and limestone, are formed from material (sediments) deposited by water or wind on the Earth's surface. Except for limestones, the material forming sedimentary rocks usually comes from the weathering and erosion of older rocks.

All rocks exposed on the surface of the Earth will eventually be weathered, eroded and re-cycled. As the top layers are removed the lower ones are exposed and meet the same fate. Eventually, by this process, rocks that were once deeply buried will be brought to the surface. This is called the Rock Cycle and operates over timescales of hundreds of millions of years.

Sedimentary rocks

All of the rocks exposed along the Beacons Way are sedimentary rocks. Sediments are first deposited as layers of loose material in a subsiding basin. The older layers are slowly buried under the younger ones and the water they contain is squeezed out as the sediments compact. Crystals grow in the tiny spaces between the grains and cement the grains together to form a rock. All this takes many millions of years.

The nature of the rock that is finally formed depends on many things – the source rock(s), the conditions of weathering and erosion and where the grains are deposited. A river fed by occasional torrential rain and rushing down a steep mountainside will transport boulders, cobbles and pebbles (which will become rounded as they collide with each other) and deposit them as soon as the slope levels out. This deposit will eventually form a conglomerate rock. There are many examples of conglomerates along the Beacons Way, both in the Old Red Sandstone and in the Twrch Sandstone. Fine mud grains will be carried in suspension far out to sea or over river plains in floodwaters when rivers break their banks, eventually sinking and forming mudstone.

The fact that the environment changes, either abruptly or gradually, means that different types of sediment are laid down in succession; and that we can deduce what those environments were like and how they changed over time from the deposits they left.

Sometimes earth movements uplift and tilt the layers and erosion wears them down before newer sediments are deposited on top. The boundary between the old and new layers is called an unconformity and usually represents a long period of uplift and erosion. There is such an unconformity between the Lower and Upper Old Red Sandstone but the difference in the tilt of the beds is so slight this unconformity is difficult to appreciate.

Fossils

A fossil is any evidence of ancient life preserved within sedimentary layers. Body fossils are the bodily remains of animals and plants, or the impression of those remains. Trace fossils are evidence of the activity of the animals, such as their tracks or burrows. There are examples of burrows in the blocks forming the steps up to Craig y Fan Ddu on Day 4.

Fossils are rare in sandstones, especially those deposited on land, as sand is too coarse a medium in which to preserve delicate structures and is usually deposited by fast flowing water so the remains are broken up and washed away. Mudstones are a better preservative and many of the mudstones within the Old Red Sandstone contain fossils of some of the oldest land plants, but they are difficult to spot.

Most limestones are actually formed of shell and coral fragments but often these have been ground up so finely that the rock is rather uniform and grey. However some limestone beds, such as at Ogof Ffynnon Ddu, contain many intact shells and even complete coral colonies.

Plate tectonics, continental drift and continental collision

The Earth is restless. The surface is not an unbroken sphere but is made up of separate plates (seven huge and eight smaller ones at present). Some of these plates are composed entirely of thin but dense oceanic crust, some of thicker, lighter continental crust and others of both types. The Earth's mantle is not completely solid, it is heated by internal radioactivity and by some residual heat from the formation of the Earth, and is ductile, especially the layer called the asthenosphere on which the tectonic plates 'float' and move, typically at about 4 cm a year (the speed at which fingernails grow). 4 cm is not much but over 100 million years this adds up to a distance of 4000 km, which is a pretty substantial drift. South Wales has drifted from near the South Pole to its present position over 470 million years and during this time its position relative to other continents has changed dramatically.

New oceanic crust is created where two plates are moving apart, as at the Mid Atlantic Ridge. Where oceanic and continental crust is colliding, as around the rim of the Pacific, dense oceanic crust is being pulled underneath lighter continental crust and subducted back into the mantle. Subduction causes volcanoes and earthquakes. When two areas of continental crust collide one cannot subduct far beneath the other as both are the same density; so the relentless movements of the plates force them together. Often the continental rocks are folded, fractured and thrust over each other to form a mountain chain on greatly thickened continental crust. This mountain building process is called an orogeny. Sometimes the collision is relatively gentle and the two plates dock quietly against each other and do not suffer significant deformation. Continental collisions are rarely simple and straightforward. Frequently the plates approach each other obliquely and may slide past each other for hundreds of kilometres, often slicing slivers off each other which then get dragged far from their origins, creating jigsaws for geologists to solve. All of these things happened when the plates carrying Avalonia, Baltica and Laurentia collided in stages over about 150 million years, resulting in the Caledonian Orogeny and the Caledonian Mountains. The mountains were then eroded to provide some of the raw materials for the Black Mountains, Brecon Beacons and Y Mynydd Ddu.

Faults and disturbances

The earth movements of plate tectonics frequently fracture the brittle crustal rocks and cause the rocks on one side of the fracture to move relative to those on the other. These fractures are usually sloping rather than vertical. If the crust is being stretched then the rock mass above the oblique fault plane slides down the plane so this side of the fault drops downwards and younger rocks now lie adjacent to older ones; this is called a normal fault. If on the other hand, the crust is being compressed, the rocks above the fault plane are pushed upwards and older rocks are forced over younger ones; this is a reverse fault. Sometimes the rocks are pushed laterally past each other with little or no vertical movement. This is a strike-slip fault, the most infamous example being the San Andreas Fault. Nearer to home the Neath, Tawe Valley and Carreg Cennen Disturbances all include components of strike-slip faulting.

A disturbance is the name given to a region where there are a number of faults which have moved in different directions at different times over hundreds of millions of years. The movements may also cause the rocks to buckle and fold. The upfolds are termed anticlines and the downfolds synclines. Disturbances are areas of weakness in the crust which tend to move and take up the strain whenever the plates move. South Wales is dissected by a number of disturbances.

Supercontinents

Around 300 million years ago plate tectonics and continental drift brought all the Earth's landmasses together to form the most recent supercontinent, Pangaea ('entire earth'), with South Wales in a desert region near its centre. Pangaea started to break up about 200 million years ago and now the continents are separate entities again. There may have been six or seven such supercontinents since the Earth originally formed 4.6 billion years ago, one approximately every 300 to 500 million years. This coming together of all the continents only to split and re-merge in a new pattern seems to be a cyclical event. The next supercontinent is due between 50 and 300 million years from now, and even has a name – Amasia, but that need not concern us, and will probably not concern any humans.

Recent sediments

During the present Ice Age, which only started 2.6 million years ago, ice sheets and glaciers have advanced and retreated across South Wales many times. The powerful erosive action of ice scraping over bedrock ground up huge amounts of this rock into an unconsolidated sediment we call till, a mixture of clay, sand, pebbles and small boulders. When the ice finally melted this till remained as a thin covering over much of the lower lying land. In places where rivers were gushing from the melting ice the clay would be washed away leaving the coarser deposits. All these glacial deposits are called drift and often hide the bedrock underneath.

The Land of the Beacons Way

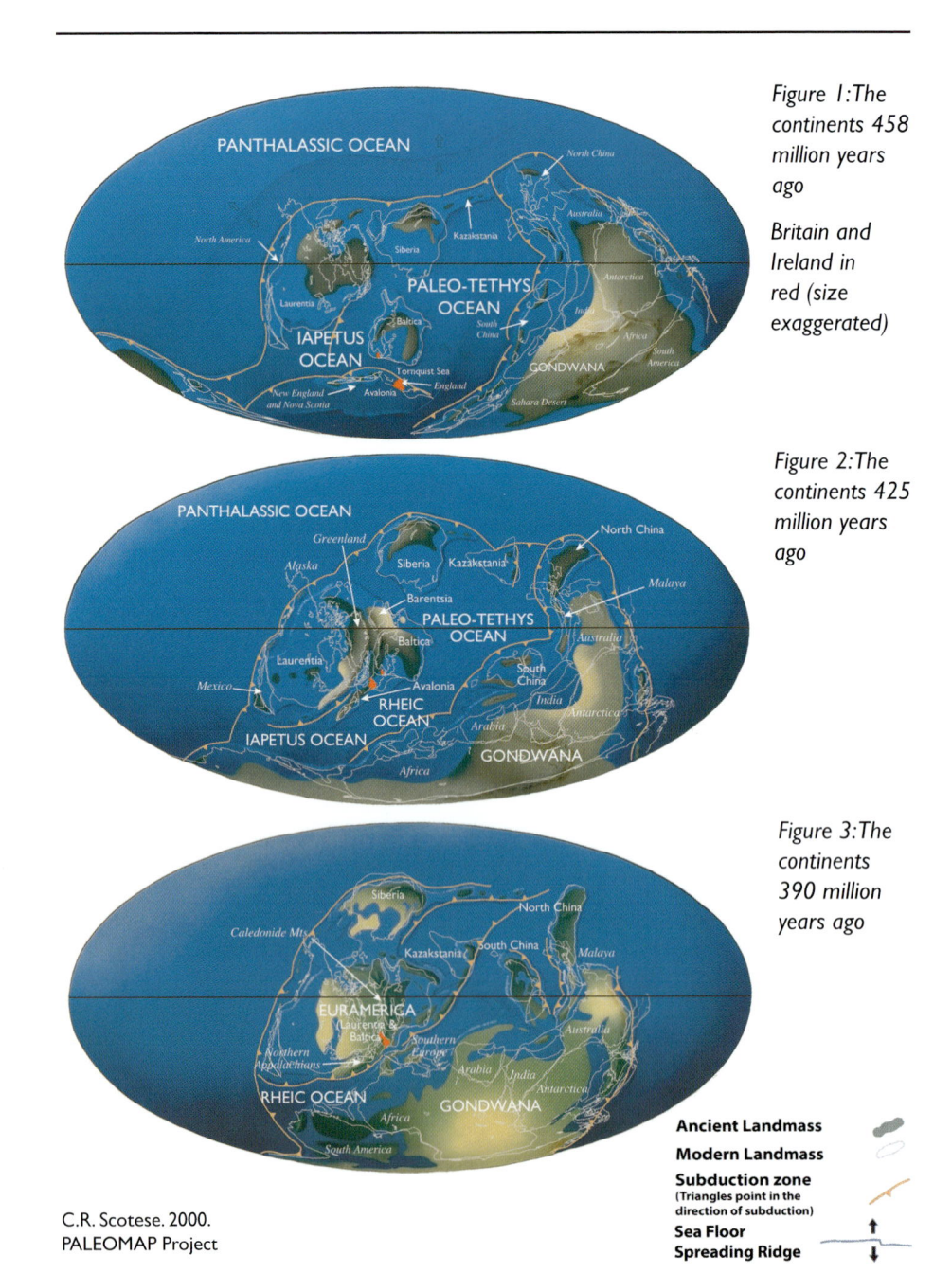

Figure 1: The continents 458 million years ago

Britain and Ireland in red (size exaggerated)

Figure 2: The continents 425 million years ago

Figure 3: The continents 390 million years ago

Ancient Landmass
Modern Landmass
Subduction zone
(Triangles point in the direction of subduction)
Sea Floor
Spreading Ridge

C.R. Scotese. 2000.
PALEOMAP Project

Part 1. The big picture
Cambrian, Ordovician and Silurian Periods, 541 – 419 million years ago
Continental collision and mountain building

At the beginning of this time southern Britain, including Wales, was part of a continental fragment called East Avalonia which lay towards the South Pole, attached to the edge of Gondwana. Gondwana was the great southern continent mainly made up of South America, Africa, Australia, Antarctica and India. The Brecon Beacons National Park lay on the southern edge of a deep sea basin, now called the Welsh Basin, on the northern margin of East Avalonia. Large volumes of mud and sand poured into the Welsh Basin, eroded from the surrounding land. These were buried, compressed and cemented producing the Ordovician and Silurian sandstones and mudstones you will cross just before Bethlehem, the oldest rocks along the Way. Later they were folded and faulted into complex structures when the Welsh Basin was squeezed and its contents forced upwards during the Earth movements and continental collisions which followed.

In the Ordovician Period, about 475 million years ago, the tectonic plate carrying East Avalonia split from Gondwana and continued to move north by the process known as continental drift towards a much larger continental block called Laurentia (North America, Greenland, Scotland and the northern part of Ireland) whose southern edge lay about 30 degrees south of the equator (30°S). It took about 50 million years for these continents to start colliding. Laurentia was also moving slowly north but East Avalonia moved faster and caught up. The two began to physically collide in the Silurian Period, around 425 million years ago, at about 25°S. By this time both East Avalonia and Laurentia had each separately collided with another continent called Baltica (Scandinavia and Russia west of the Urals). The lengthy mountain building episode associated with the complex collision of East Avalonia, Laurentia and Baltica is called the Caledonian Orogeny. This lasted until the middle of the Devonian Period, and created the continent of Laurussia aka Euramerica (the Old Red Sandstone Continent) with the now united Britain and Ireland on its southern edge. The Iapetus Ocean which once separated the three continental blocks was subducted under them and completely disappeared. The suture line where Laurentian Scotland and Avalonian England and Wales were welded together runs roughly north-east from the modern Solway Firth and south-west through the Isle of Man into Ireland.

Many of the rocks caught up in the Caledonian Orogeny were pushed upwards, folded, thrust over each other, heated up and intruded with granites, to form mountains which were probably as high as the Alps. This Caledonian Mountain chain stretched right across Norway, northern Britain, Newfoundland, Canada and the Atlantic coast of the

USA, were now part of the one landmass, Laurussia, still moving slowly northwards.

Recycling the Caledonian Mountains

Even while they were still rising the Caledonian Mountains were being eroded. The only land vegetation was surface mats of algae and bacteria with the very earliest small vascular plants in damp niches. There were no forests to protect the new mountains from either the baking sun or from snow and ice on their summits, so the rocks were rapidly weathered and frost shattered and the debris washed downwards by heavy rains. Enormous quantities of rock were eroded off the mountains, ground up and carried, mainly southwards by great rivers to be re-deposited as the Old Red Sandstone (ORS). These rivers were nothing like modern Welsh rivers. Imagine instead the powerful rivers flowing off the Himalayas onto the Indian and Pakistani plains today. As the waters left the mountains and uplifted areas they slowed down and spread out into huge meandering or braided rivers which flowed across a wide coastal plain to the ocean which then stretched southwards from Devon. Initially, during Silurian times, the debris reaching South Wales very probably came from the Caledonian Mountains themselves, but later this debris was trapped in sedimentary basins forming further to the north and the ORS deposited across South Wales was sourced from nearer uplifted areas in northern Britain and Wales.

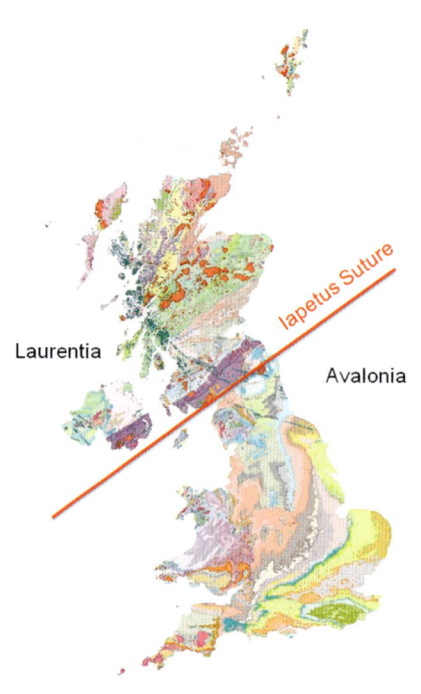

Reproduced with the permission of the British Geological Survey ©NERC. All rights Reserved

Figure 4: The Iapetus Suture

Figure 5: A braided river, the Rakaia, South Island New Zealand

Devonian Period, 419 – 359 million years ago

By the Devonian Period, Britain and Ireland had drifted into the semi-arid subtropics at about 15 - 20°S. South Wales, in common with most of the adjacent land, was a desert; a great red plain seasonally crossed by ever shifting braided river channels which occasionally burst their banks and flooded widely, leaving ephemeral lakes before they too dried up.

In the early part of the Devonian Period the crust of Britain was being stretched by plate tectonic movements to create shallow basins, including one across the southern part of Wales, south of the old marine Welsh Basin. The eroded debris accumulated in this basin as horizontal layers and as the sediment layers thickened the basin sagged under their weight, making room for more layers.

The weight of the rising Caledonian Mountains to the north may also have depressed the crust beneath the basin. The Black Mountains, Brecon Beacons and Y Mynydd Du (the Black Mountain) are all made of these ORS sedimentary layers. By the end of the Devonian Period the great mountains formed in the Silurian Period had been almost flattened by erosion and recycled southwards.

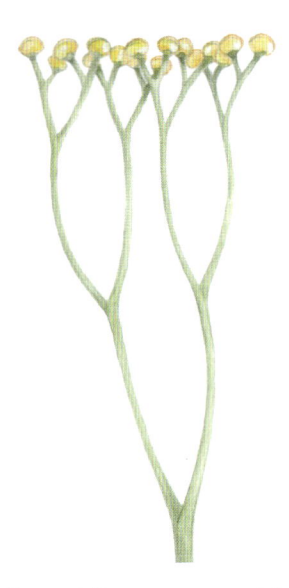

Life moves onto the land

During the 60 million years of the Devonian Period terrestrial plants evolved from the tiny, simple forms with no root systems which had originated during the Silurian Period, into a wealth of larger plants and trees (but there were no flowering plants). Many different sorts of fish evolved, the ancestors of the first four legged animals appeared and vertebrates moved onto the land. Fossils of some of these earliest land plants and freshwater fish have been found in the rocks of the Brecon Beacons National Park together with animal burrows and trackways. Animals and plants were appearing on land during the Devonian Period but today these rocks in the Park seem devoid of fossils because preservation was rare in this terrestrial, high energy environment.

Figure 6: Reconstruction of the early vascular plant Cooksonia showing naked stems ending in spore sacs. This plant was only a few millimetres high

Floods and droughts

During storms and flash floods the rivers carried boulders and pebbles which were deposited as soon as the gradient of the river bed or the volume of water decreased. These formed conglomerate rocks. The river channels transported sand grains of various sizes and ripple marks formed as currents moved the grains along. Where the rivers flowed into lakes or burst their banks and flooded the plains, finer sand, silt and mud were deposited as the water velocity suddenly dropped. When the floods subsided or the lakes dried up the mud surfaces would crack, some of these cracks are still visible. The ORS sediments contained some lime (calcium carbonate); the occasional rains would dissolve this lime and wash it downwards through the sediments, then the intense heat would evaporate the water, drawing the lime upwards to be precipitated in the top layers. After many repetitions of these wet-dry cycles the calcium carbonate became concentrated into hard bands of knobbly limestone just below the land surface. These bands are called calcretes and represent fossilised desert soils. Each calcrete also represents about ten thousand years when no fresh sediment was being deposited. Modern calcretes are forming in hot desert areas today. A modern-day analogue for the Lower Old Red Sandstone environment is the Channel Country in central Australia.

Four Formations

Geologists find it convenient to divide the Lower ORS into four formations:

1. The lowest (oldest), the Raglan Mudstone Formation, was actually laid down during the Silurian Period and is predominantly mudstone from a river and mudflat environment.

2. The St Maughans Formation comprises the deposits of sluggish meandering, probably ephemeral, river channels, plus thick muddy floodplain and lake sediments, frequently containing calcretes. Deposition followed a cyclical pattern. The start of each cycle was marked by an erosion surface cut into the underlying mudstone by the new river channel. Next an intraformational conglomerate mostly consisting of pieces ripped up from the older mudstones and calcretes was deposited in the bottom of the channel, followed by sandstone fining upwards into siltstone. When the river flooded the channel was smothered by mud which often eventually dried out, cracked and became calcretised, so completing that cycle. The St Maughans Formation is about 65% sandstone and 35% mudstone

3. The third ORS unit, the Senni Formation, is mostly sand carried by seasonally flowing braided rivers. Another modern equivalent would be rivers like the Indus carrying millions of tons of sediment from the Karakoram Mountains onto the semi-desert plains of Pakistsan and occasionally causing widespread flooding, as in 2010. Surprisingly it was in this rather hostile environment that land plants really took off, flourished and diversified.

4. The uppermost (youngest) division is the Brownstones Formation, mainly sand and pebbles

Figure 7: Conglomerate in the Brownstones Formation

Figure 8: Ripple marks on the summit of Pen y Fan

Figure 9: A modern calcrete forming in southern Australia

deposited in networks of braided channels during flash floods. Many of these channels must have been huge as some sandstone beds are kilometres in width. There are almost no plant or animal fossils in the Brownstones Formation, but that does not mean this time was lifeless, far from it, just that conditions were not right for their preservation.

Much of the Old Red Sandstone is not sand and some of it is not even red. The upper divisions of the ORS are more sandy and pebbly and less muddy than the lower ones, because the mountainous source areas were nearer and the rivers more energetic.

Figure 10: The Indus river carrying debris from rapidly eroding mountains

In today's landscape the softer muddier layers have readily weathered and eroded, tending to form low lying ground. The harder sandy or pebbly beds usually form the higher hilltops and escarpment faces. Alternations of hard and soft beds produce a stepped appearance in the landscape, classically displayed around Pen Cerrig-calch. They also give rise to small waterfalls along the escarpments as at Fan Hir.

Figure 11: The stepped hillside of Pen Cerrig-calch, looking north from Sugar Loaf

Red rocks

Many of the ORS rocks are red because the sand and mud grains have a thin coating of iron which was oxidised to form the mineral haematite (Fe_2O_3) in the hot, dry air. Some ORS rocks contain green bands or spots. These form where heamatite has been altered due to the removal of oxygen by decaying plants or animals. The Senni Formation is generally greenish due to the presence of the mineral chlorite which forms from mica in the sandstones, probably because the water table was higher during this time. This high water table was probably the reason why plant remains were better preserved in this formation.

Figure 12: The top of Pen y Fan showing the Plateau Beds (upper few metres) above the sheet sandstones of the Brownstones

Another collision?

Around 390 million years ago, towards the end of the Lower Devonian, more plate tectonic movements and a possible further collision along the southern edge of Laurussia uplifted the Welsh ORS alluvial plain. Sediments typically do not accumulate on uplifted areas so no deposits from the Middle Devonian Epoch have been identified in South Wales and there is an unconformity (time gap) of about 20 million years between the Lower and Upper ORS across the National Park. In addition, a greater than 3 km thickness of the already deposited Lower ORS was stripped off and removed during this period of uplift. So we now only see a fraction of the original Lower ORS.

The rest was swept out to sea and recycled somewhere else. Another result of this collision was that the Ordovician and Silurian sediments in the Welsh Basin were compressed, folded, faulted and pushed upwards. It was this compression which converted many of the mudstones into slates, thereby providing high quality roofing for much of the Britain and Ireland. This mountain building episode is called the Acadian Orogeny and geologists are still trying to identify the responsible event. It is possible that something in the Rheic Ocean, which had opened up to the south of the Old Red Sandstone Continent, collided with East Avalonia. That something could have been a young, buoyant section of ocean crust or the continental fragment of Iberia which had also separated from Gondwana and was moving northwards.

The Land of the Beacons Way

Figure 13: Plateau Beds above Llyn y Fan Fach

Figure 14: Sugar Loaf with a hard flat top of Quartz Conglomerate

Sea levels rise

During the following Late Devonian Epoch the sea level rose, partly because the compression of the Acadian Orogeny was waning so the land was no longer rising, and partly because an icecap over the South Pole was melting. South Wales now lay on the northern edge of an encroaching sea and braided rivers again swept down from the mountains rejuvenated by the orogeny, depositing sand and pebbles on the coastal plain. All the resulting Upper ORS rocks are hard and frequently pebbly and originally formed extensive plateaus, the remnants of which are now protective caps on many summits from the Y Mynydd Du (the Black Mountain) eastwards. The Plateau Beds Formation which tops Bannau Sir Gaer, Fan Foel, Fan Brycheiniog, Fan Gyhirych, Pen y Fan and Corn Du, contains marine fossils and are thought to have accumulated on a tidal flat. The Quartz Conglomerate Group on Sugar Loaf, Pen Cerrig-calch and Table Mountain was probably deposited at the seaward edge of an alluvial fan.

Carboniferous Period, 359 – 299 million years ago

Sunlit tropical seas

The tectonic plates were now colliding more slowly so the pressures were relaxed and the rocks cooled. South Wales sagged and was flooded by a shallow sea with coastal lagoons, like the Bahamas today. This was the start of the Carboniferous Period and South Wales

Figure 15: Siphonodendron, a colonial coral in limestone, Ogof Ffynnon Ddu

Figure 16a: A diagram of a crinoid

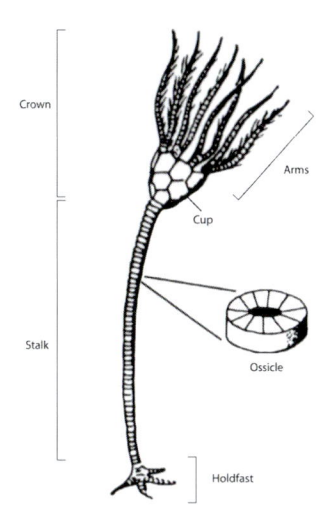

Figure 16b: A fossil crinoid

lay just south of the equator. The climate was still semi-arid with few rivers and the land was low-lying, so little sand and mud was washed in to cloud the clear tropical sea. Corals, crinoids (animals related to sea urchins but called sea lilies) and many types of shelled creatures flourished. Their calcium carbonate remains were cemented to form limestone (Rocks of this age are formally referred to as the Avon Group and the Pembroke Limestone Group). Sometimes the corals and shells are visible in the rocks but often they have been finely ground into a lime mud, making a rather uniform grey rock. Where conditions were rough, as on bars and in channels at the seaward margins of lagoons, shoals of ooliths formed. Ooliths are round grains, 0.5 – 2 mm in diameter, formed when fragments of shell are rolled around in water saturated with calcium carbonate, which then precipitates as concentric layers around the fragment.

Sea levels fell occasionally (as ice started to form over the South Pole, see below), exposing the limey sea bed which dried out, was colonised by plants and developed a thin soil cover. These surfaces, now called

Figure 17: An exposed paleokarst surface

The Land of the Beacons Way

palaeokarsts, are fossilised limestone pavements. This idyllic state of affairs lasted for 30 million years.

Cooling climate, deltas and grit

Halfway across the world, where the southern edge of Gondwana lay over the South Pole, major ice sheets were forming in the Middle Carboniferous (Namurian) Period 325 million years ago. One reason proposed for this glaciation was the removal of huge amounts of the greenhouse gas carbon dioxide from the atmosphere, so cooling the world. The formation of limestone from carbonate skeletons locked away millions of tons of carbon dioxide, and the burgeoning Carboniferous forests sucked up even more and stored it as peat (subsequently to become coal). As the ice sheets grew sea levels fell, South Wales became low lying land again and great thicknesses of Carboniferous limestone, were stripped off by erosion.

Figure 18: Twrch Sandstone in Cwm Twrch

Figure 19: Sleeper block of Twrch Sandstone, Penwyllt.

The Big Picture

Figure 20: Penwyllt Brickworks

Across mid Wales there was an area of higher land called the Wales–London–Brabant High (formerly known as St George's Land). It was the old core of East Avalonia and stretched from Ireland to Belgium. The Brecon Beacons National Park now lay on the edge of the shallow water on its southern shore, just north of the equator, with a hot, humid climate. The Wales-London-Brabant High was being eroded and rivers carried the debris southwards into a Mississippi sized delta of river channels, depositing it as pebbly or coarse sand layers known as the Twrch Sandstone (previously called the Basal Grit); or, when the delta channels burst their banks (as happened around New Orleans in 2005) in muddy layers (the Bishopston Mudstone Formation). These two formations which were originally called the Millstone Grit have been renamed the Marros Group.

Occasionally, on the beaches at the delta edges, the sand would be washed so thoroughly by the waves that it became almost pure silica sand. This formed the

Figure 21: Stigmaria in Twrch Sandstone on Garreg Las

silica rock which was crushed, shaped and fired to produce refractory bricks which were used to line blast furnaces as they withstood high temperatures. Silica rock and sand were quarried near Penwyllt, on Cribarth and on Y Mynydd Du. Bricks were manufactured at a number of sites, including the Penwyllt Brickworks (the ruins of which can still be seen at SN 855 152, just south of the Caving Club headquarters).

The Southern Hemisphere climate warmed periodically, the ice sheets melted, and the rising sea flooded over the delta tops leaving 'marine bands' of dark shale with marine fossils in the rock sequence. During cold intervals the ice sheets expanded and sea levels fell exposing the delta tops which were colonised by swampy tropical rain forests. The fossilised 'roots' of some of these trees (Stigmaria) can be seen in the Twrch Sandstone, together with fallen branches and tree trunks in the delta channels.

Tropical Coal Swamps

The Beacons Way does not actually cross the South Wales Coalfield. However the coalfield must be mentioned because it is so important to the history of our area. The occurrence of coal and iron ore deposits in the Coal Measures, together with the Namurian silica rock plus Carboniferous limestone around the margin of the coalfield drove the Industrial Revolution in South Wales.

During late Carboniferous times, when the Coal Measures were being deposited, 310 – 299 million years ago, the delta tops and their coastal plains stabilised, becoming areas of lakes, swamps and raised mires. South Wales lay on the equator; luxuriant vegetation grew rapidly because of the wet tropical climate and the high atmospheric oxygen level (about 14% more than today), itself a product of photosynthesis by the increasing biomass. Millipedes reached one metre in length and dragonflies achieved wingspans of up to 70 cm. The plants were not like modern woody trees. Instead there were giant clubmosses, horsetails and tree ferns, all with shallow anchoring systems and a soft structure so they collapsed and rotted easily. The water table was high and there was little oxygen in the stagnant water so fallen vegetation rapidly decayed in the hot anoxic conditions to form thick peat layers. When the delta channels burst their banks or sea levels rose, the

Figure 22: Coal forest swamp

The Big Picture

swamps were flooded with mud, killing the trees and ensuring a succession of rotting vegetation.

Sea level falls meant that sand was carried down the rivers, filling their valleys. The Farewell Rock at the base of the Coal Measures is such a valley-fill. These cycles were repeated many times. Over millions of years the peat was buried under successive layers of mud, more peat and sand, and compressed to form coal. A ten metre thickness of peat yielded one metre of coal. This is why most of the Coal Measures rocks are mud and sand with only relatively thin coal seams.

Supercontinent

Gondwana, the great southern continent, eventually caught up and collided with Laurussia, the continent formed by the amalgamation of North America, Greenland, Baltica, Avalonia and Armorica. This oblique collision, the Variscan Orogeny, started in the late Devonian Period and continued throughout the Carboniferous Period into the early Permian Period. It lasted for 100 million years and created Pangaea, the most recent supercontinent. The collision threw up another great mountain chain, about 7000 km in length and up to 10,000 m high, stretching from the Gulf of Mexico to Eastern Europe. These Variscan Mountains formed across mainland Europe, and the collision buckled and folded the rock layers across southern Britain.

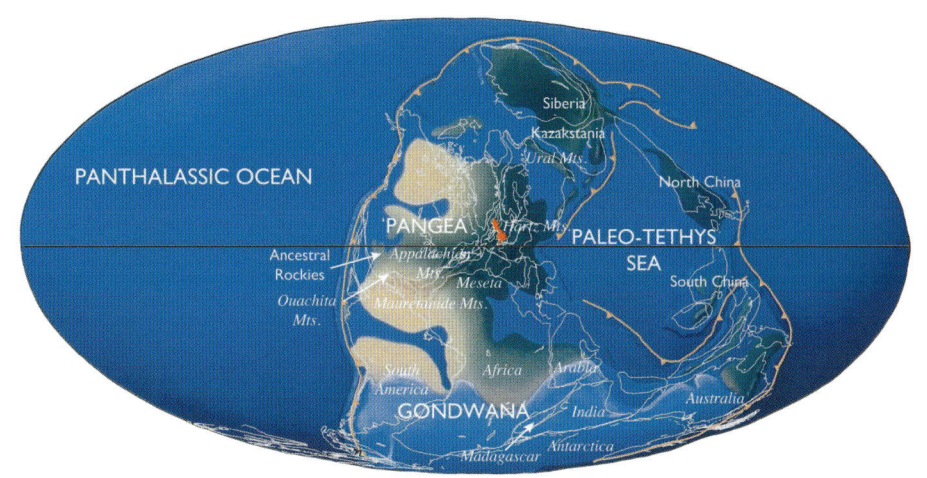

C.R. Scotese. 2000. PALEOMAP Project

Figure 23: The continents 300 million years ago (Britain and Ireland exaggerated in size)

The Land of the Beacons Way

South Wales was compressed from the south by this collision about 280 million years ago and folded into an east-west elongated bowl (syncline), the Coal Basin. Because the compression came from the south, the southern side of the bowl is much steeper than the northern side. The Brecon Beacons National Park lies on the northern rim of this bowl, so all the rock layers here now normally tilt gently to the south. Older rocks are exposed around the rim and younger rocks in the centre of the bowl. Within the bowl the rocks were also fracturing and folding during the orogeny. This is why the Carboniferous limestone, Marros Group (Millstone Grit) and Coal Measures rocks are broken up by so many folds and faults.

Ancient cracks in the crust

Extending from south-west to north-east across South Wales are a series of deep lines of weakness in the Earth's crust, manifested as zones of fractures and folds and called disturbances. Their origins are uncertain but they go back at least to the Caledonian Orogeny, maybe even earlier. These structures include, from west to east, the Carreg Cennen Disturbance, the Tawe (or Swansea Valley) and Cribarth Disturbances and the Vale of Neath Disturbance. The Carreg Cennen and Tawe/Cribarth Disturbances are arms of the Welsh Borderland Fault System, a faulted zone extending from Pembrokeshire to the Peak District, which marked the south-eastern margin of the Welsh Basin between 541 and 419 million years ago.

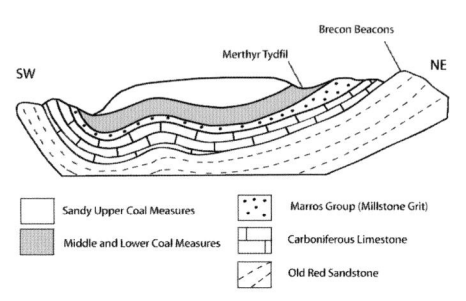

Figure 24: Diagram of the South Wales Coal Basin

Whenever South Wales was compressed or stretched there would be movement along these weaknesses, sometimes vertically, sometimes sideways slipping or both together. These disturbances are still active because the tectonic plates are still moving and the Atlantic Ocean is still widening. Earthquakes have been recorded in Swansea approximately every 50 – 100 years since at least the early 1700s. The latest, in 1906, measured magnitude 5.2. toppled chimneys and wobbled the Mumbles lighthouse. Sennybridge was shaken by a 3.5 magnitude quake in 1999 and in May 2012 there was a small quake along the Neath Disturbance south of Llangynidr. Erosion along these lines of weakness has produced the Tawe (Swansea) and Neath Valleys, explaining why they are deeper and straighter and have a different trend to other South Wales valleys.

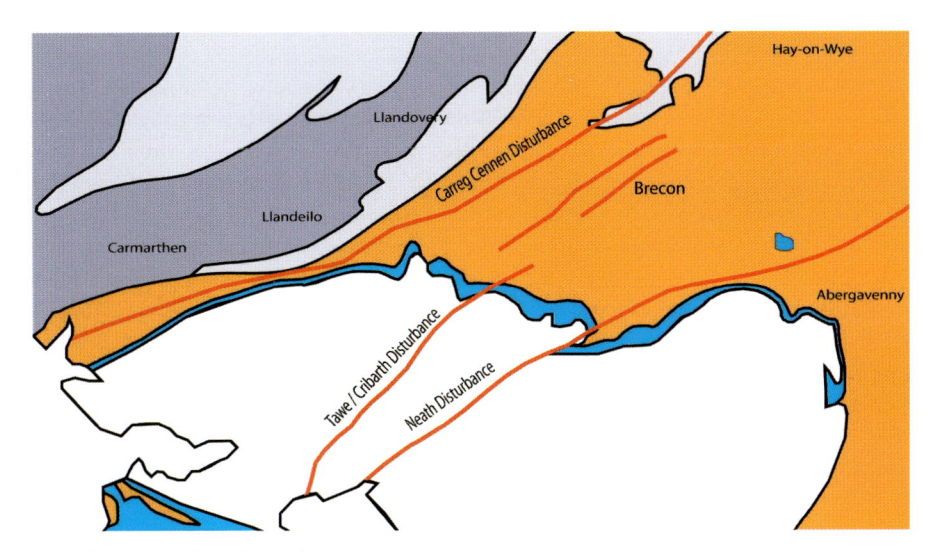

Figure 25: Map of the Disturbances

Permian, Triassic, Jurassic, Cretaceous and Tertiary Periods, 299 – 2.6 million years ago

(The Tertiary Period has now been re-named the Palaeogene and Neogene, 66 – 2.6 Ma)

Pangaea

By the end of the Variscan Orogeny all the Earth's landmasses were welded together into the supercontinent Pangaea. Britain and Ireland were roughly in the centre, in the northern desert belt where the Sahara lies today. We were an upland desert area again and great thicknesses of the Coal Measures must have been stripped off by erosion. What we see now is a tiny fraction of what was originally there.

By 180 million years ago the supercontinent started to split violently apart and the southern Atlantic began to open, separating Laurussia (now part of Laurasia) and Gondwana again.

The missing millions

No rocks deposited during the last 300 million years are now found in the Brecon Beacons National Park. During this time Wales began as a desert with ephemeral lakes (Permo-Triassic); became an island archipelago in a sub-tropical sea when sea-levels rose as Pangaea split apart during the Jurassic (245 – 145 million years ago); may have been completely flooded by a deep, warm Cretaceous sea (145 – 66 million years ago) and then was probably uplifted and eroded again. Sediments such as the thick Cretaceous Chalk, may have been laid down, but if so, they too were stripped off by erosion. By about 50 million years ago South Wales was probably a low lying, featureless plateau, tilted gently to the south. The Beacons Way would have been a very different walk. During all this time our tectonic plate had been moving slowly northwards and by 50 million years ago Wales was approximately in its present position.

Hot and wet

About 55 million years ago temperatures rose worldwide, probably by as much as 6°C over 20,000 years. The reason(s) are not known but it is thought that there was a massive release of carbon into the atmosphere and oceans. The climate also seems to have become much wetter. The result in our area was the deep tropical-style weathering which disaggregated areas of Twrch Sandstone producing the easily extracted silica sand used for furnace bricks during the industrial revolution. The Way passes several areas of this silica sand between Llanddeusant and Carreg Cennen.

Up again and levelled off

The whole of Wales was uplifted by at least 1000 metres sometime between 66 and 2 million years ago, possibly in a series of pulses. These pulses of uplift may have created domed land surfaces at differing heights above sea level, determined by the underlying geology. The domes were then eroded and flattened by rain, rivers and fluctuating sea levels, leaving the succession of almost flat surfaces at different heights which are obvious when you look across the landscape. Two events may have been responsible for the uplift. One was the opening of the North Atlantic when, about 63 million years ago, a hot spot and mantle plume (now under Iceland) made the Earth's crust dome upwards and split apart just to the west of the British Isles. The second was another plate collision, the Alpine Orogeny. This time the African plate swung slowly round and northwards to collide with Europe and throw up the Alps between 10 and 15 million years ago. There was no associated mountain building in the British Isles but the land may have been pushed gently upwards.

The strata in our area still generally dipped gently southwards because we were on the northern side of the South Wales Coal Basin. Rivers continued to cut down into the underlying rocks and slowly established the broad outlines of our landscape.

Quaternary Period, 2.6 million years ago to the present

Ice

It was ice that put the finishing touches to the landscape and carved today's mountains, scarps & valleys. Since the Ice Age began, 2.6 million years ago, most of the British Isles is thought to have been glaciated at least eight times. Ice masses ranging in size from small glaciers to big ice sheets repeatedly advanced and retreated. The ice built up slowly over thousands of years. The critical factor was the summer temperature; if this was too low to melt the previous winter's snowpack, successive snowpack layers built up and were compressed into ice by their own weight. The Ice Age's cause is uncertain, probably many interacting factors contributed.

Sculpture by ice and water

Figure: 26: The two arms of Virkisjökull and Falljökull separate and then recombine around the nunatak Raudikambur in Iceland

The ice flowed slowly from the highlands to the lowlands under the influence of gravity. It shattered and scoured the rocks, gouged out lines of weakness (especially along faults) to create valleys, then deepened and widened these valleys. Rivers could run underneath the thick ice; this water was under high pressure and could force rocks apart. When the climate warmed the ice melted rapidly, quite possibly in as little as 10 – 100 years. Great volumes of water were liberated, cutting gorges through hard rock or spreading sideways as braided rivers over softer sediments. Lakes formed in overdeepened areas or where the water outlet was blocked either by ice or glacial deposits, as at Llangors.

Dumps

Glacier ice contains a lot of rock debris, either plucked out by the moving ice or as frost-shattered blocks fallen from above. Most of this debris was ground up into till (formerly called boulder clay), a jumbled mixture of clay, sand, gravel and boulders which was carried along within the ice. When the ice thawed the glaciers usually retreated step-wise up the valleys, dumping banks of till, called moraines, where they paused. There is a splendid example near Llanfihangel Crucorney. Moraine ridges also formed along the glaciers' sides because friction makes the ice move more slowly at the sides than in the centre. More till was smeared over most of the land beneath the ice and forms the basis of the present soil cover. Glaciers also carried large boulders for many miles and left them dotted over the land surface as erratic blocks when the ice melted. Tracing the source of erratics helps to track the directions of ice movement.

Figure 27: ORS erratic perched on a gritstone pedestal

Limestone pavements

Moving ice removed the soil and weaker rock strata covering the strong beds of Carboniferous limestone, leaving them smooth but covered with a layer of till. As a vegetation cover grew on the till the roots produced humic acids which, together with acidic rain-water, seeped down to dissolve the limestonel along joint planes to produce vertical fissures in the rock, known as grikes. The grikes grew wider, allowing the soil

The Big Picture

Figure 28: Limestone pavement, Ogof Ffynnon Ddu

cover to be washed away; more recently clearance and grazing have accelerated this process. Unlike other rocks, pure limestone does not weather into soil, so eventually only a bare surface of limestone pavement remained with raised blocks (clints) between the grikes. The limestone pavements in the South Wales are smaller and more shattered than those in places like the Yorkshire Dales, because the Carboniferous limestone here was fractured and folded during the Variscan Orogeny.

Shake, swallow and sink

Although Carboniferous limestone is mechanically very strong, it dissolves away leaving fissures, caves and tunnels, often large and extensive. The distinctive landscape formed on soluble rocks with an efficient underground drainage is known as karst. In this landscape valleys with streams are rare. Their place is taken by dolines, depressions where water sinks underground.

Solution dolines form on exposed limestone as rain water dissolves the rock and then drains away through fissures at the bottom of the funnel shaped holes. Suffosion dolines form entirely within the overlying loose material, which slowly washes down the underlying fissures.

Caprock dolines occur when the overlying insoluble rocks collapse into an underground void in soluble strata. They are the commonest feature where beds of sandstone and grit overlie limestone and are usually called shakeholes in the UK. Sinkholes form where a stream flowing over insoluble rocks meets fissured limestone and sinks underground and so have water flowing into them, at least at times. In practice most dolines form by a combination of these events and the large examples marked on the OS maps are usually now caprock dolines.

Active sinkhole (swallow hole)

Solution doline

Caprock doline

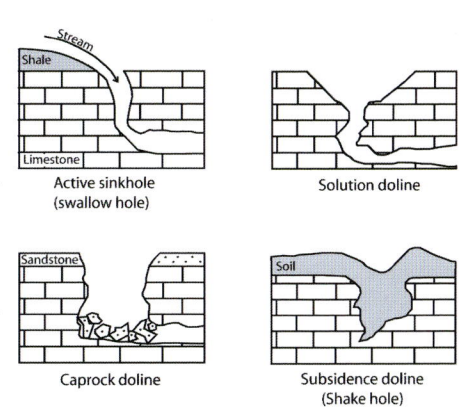

Subsidence doline (Shake hole)

Figure 29: diagram of dolines

29

Figure 30: Karst with dolines, Bannau Sir Gaer

These solution processes, which probably started in the late Neogene (23 – 2.3 million years ago), have been most active throughout the Quaternary and are still continuing. Karst with dolines occurs on Ogof Ffynnon Ddu National Nature Reserve and the lower slopes of Bannau Sir Gaer. By the time it emerges from Ogof Ffynnon Ddu, the stream which entered the ground at Pwll Byfre has traversed the deepest and one of the longest cave systems in the British Isles.

Interglacial periods

During these periods, which each probably lasted more than ten thousand years, plants spread north from refuge areas in southern Europe, Wales would have been densely forested with temperatures often 2 – 3°C higher than today. Briefer temperate interludes called interstadials also punctuated the glacial periods. In one of these, 33,000 years ago, modern humans are known to have lived on the Gower (the 'Red Lady' of Paviland has recently been re-dated).

The last ice sheets

Each ice advance removed the majority of the evidence for the previous one. So most of the features and deposits we now see in the Brecon Beacons National Park date from the last big glaciation, the Late Devensian. This got under way 26,000 years ago, reached its peak about 20,000 years ago and was ending 15,000 years ago. Late Devensian ice accumulated over the Brecon Beacons. North of Storey Arms, this flowed north and north-east to meet and feed a major glacier flowing east along the Usk Valley to just beyond Abergavenny. Ice over the Cambrian Mountains formed the central Wales Ice Sheet. Some of this flowed south-east around Mynydd Epynt into the middle Wye Valley and the Llangors area. Ice also overtopped Mynydd Epynt and augmented the Usk valley glacier. Ice over Y Mynydd Du flowed south down the Tawe Valley.

Figure 31: Moraine at Llyn y Fan Fach

A cold snap

Although the main glaciations ended 15,000 years ago, there was a brief cold period in north-west Europe from around 12,500 to 11,500 years ago, possibly due to shut down of the Gulf Stream. This is called the Younger Dryas Stadial. In Wales the forests died and were replaced by plants of the arctic tundra. Small, high level glaciers formed, especially in north-east facing cwms with large high plateaus to their west on which snow could collect and from where it could be blown into the cwms by the prevailing wind. These cirque glaciers ground out and enlarged the cwms and finally dumped moraines across their outlets. These moraines are, with the exception of landslips, the youngest and least altered landforms in the area. Good examples along the walk are at Llyn Cwm Llwch, Craig Cerrig-gleisiad and Llyn y Fan Fach.

Landslips

When the ground was frozen any valley sides and escarpments undercut and oversteepened by glacial scouring stayed hard and stable. But when the thaws came many became unstable, causing rockfalls and landslips. There are many landslips to be seen along the Way. The largest are the west flank of Ysgyryd Fawr, Crug Hywel (Table Mountain) and Darren below Pen Cerrig-calch, the western headwall of Craig Cerrig-gleisiad, the eastern slope of Fan Dringarth and at Cwmyoy; the latter is monitored and

The Land of the Beacons Way

Figure 32 Darren landslip on Pen Cerrig-calch

still moving very slowly. It now seems that some of these slips may be due to long-term weathering and weakening of the surfaces between the layers of rock, rather than just to ice.

The rest is history

The story is not just about rocks; people have used the land and its resources for thousands of years and in so doing have modified the landscape. 7000 years ago Stone Age people started to clear the forests. Once the trees were gone erosion washed away much of the mountain soil. What was left became waterlogged, especially when the climate deteriorated, so blanket peat bogs developed. Neolithic and Bronze Age folk left their stone monuments on high land. Iron Age people worked iron ore and constructed settlements and hillforts. The Romans built camps and roads, mined metals and quarried limestone for mortar and plaster. Limestone has also been quarried and burnt, first with anything available, then charcoal and later with coal, to improve fields since the Middle Ages. The ORS and other formations have been quarried extensively over the centuries for building stone and roofing tiles.

The Industrial Revolution demanded ironstone, coal, silica for furnace bricks, rottenstone for polishing metal and limestone, the latter both as a flux in the blast furnaces and also for agricultural use to help feed the increasing local population. A network of tramroads, canals and, later,

Figure 33: Limekilns at Penwyllt

The Big Picture

railways and roads was built to transport these raw materials. Around these resources, industries and transport routes communities grew up, flourished and declined leaving domestic and industrial ruins. More recently limestone has been, and still is, quarried for roadstone and aggregate. One of the Penwyllt quarries reopened in 2007 to provide aggregate for the latest gas pipeline.

Modern communities have developed needing communications, utilities and leisure opportunities. Agriculture and forestry, themselves dependent on the underlying geology and the climate, together with tourism (including walking), continue to modify the landscape. Weather is an active force; in the winter of 2009 repeated freezing and thawing followed by heavy rains caused large landslips scarring the sides of Cwm Llwch. What we see as we walk is a landscape which has evolved over 470 million years, and is still evolving, a dynamic process involving geology, climate, plants, animals, weather and people.

Figure 34: 2009 landslips around Cwm Llwch, also showing the moraine

33

The Land of the Beacons Way

Day 1. Ysgyryd Fawr (The Skirrid) to Llanthony
Maps: British Geological Survey 1:50,000 Sheets 232 (*Abergavenny*) and 214 (*Talgarth*)

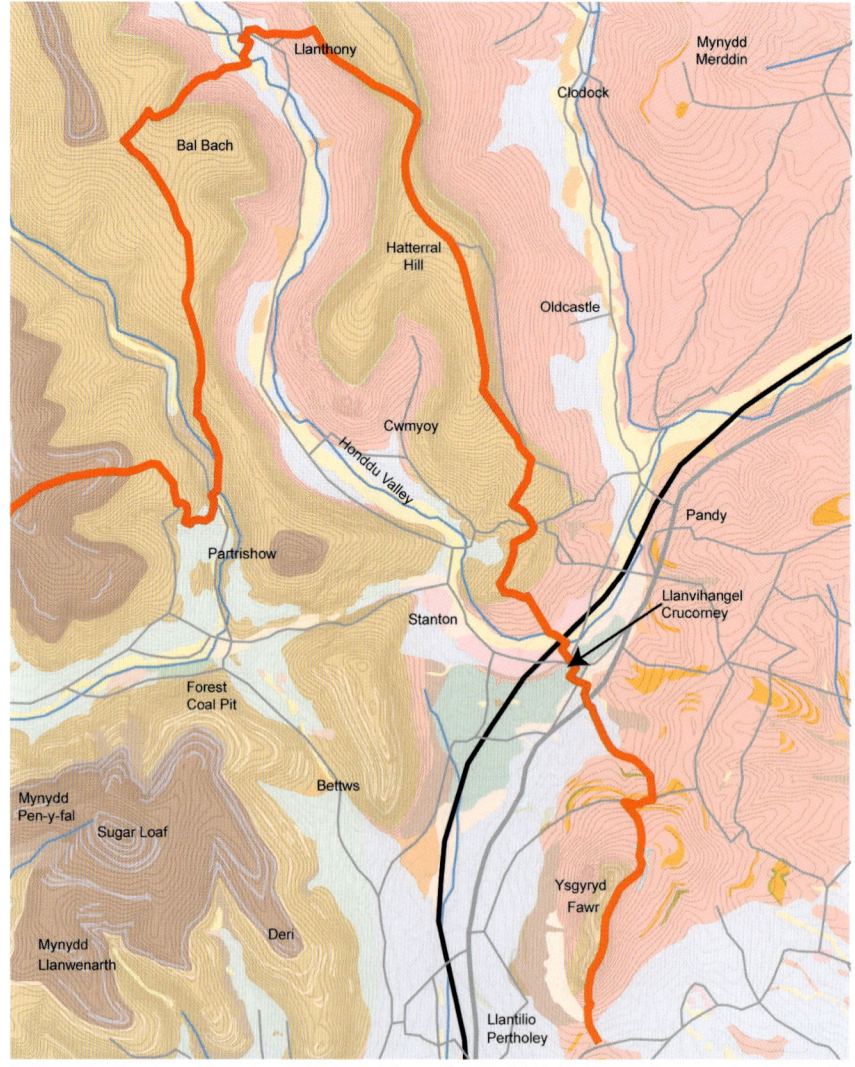

DiGMapGB50 geological data, British Geological Survey © NERC 2014. Contains Ordnance Survey data © Crown Copyright and database right 2014

Geology map for Day 1. with route marked

Highlights:
Views
Ysgyryd Fawr landslips
Llanfihangel Crucorney moraine
Llanthony Priory

Synopsis of the walk
Throughout today, the bedrock you are walking over is Old Red Sandstone (ORS), laid down in the Lower Devonian. You can see evidence for seasonally flowing, braided rivers which were prone to flash flooding, pouring off high, eroding mountains to the north-east and depositing their sediment load as sand and mud on a semi-arid coastal plain dotted with ephemeral lakes.

Ysgyryd Fawr was separated from the main plateau of the Black Mountains by river and ice erosion.

From the summit path there are wide views including to the northern rim of the South Wales Coal Basin and the Black Mountains.

The western and southern slopes of Ysgyryd Fawr are scarred by large landslides, the cause of which is still open to interpretation.

The village of Llanfihangel Crucorney is built on a glacial moraine which has a complicated history.

On your way up to Hatterrall Hill you cross the top of the valley formed along the Neath Disturbance.

Hatterrall Hill provides more exposures of ORS and excellent views.

Llanthony Priory is built of ORS and beautifully situated in a glacially formed valley.

Details:
Overview of the area
Ysgyryd Fawr is really part of the Black Mountains' upland plateau but has been separated from the main block by river and ice erosion. Between 25 and 5 million years ago relative sea levels fell steadily as South Wales was uplifted, so the Monnow and Honddu Rivers flowing south-east from the Black Mountains to join the ancient River Usk near Abergavenny were forced to cut downwards through the rock layers. Once through the harder Brownstones and Senni Formations their task was easier as the softer St Maughans Formation was more easily washed away, creating a valley. What the rivers had started, the ice probably continued and amplified. Successive glaciations destroy

The Land of the Beacons Way

evidence of the previous ones but we believe that, since the latest Ice Age began about 2.6 million years ago, successive periods of cold climatic conditions allowed long-lived snow or ice to form in the Black Mountains on a number of separate occasions. This ice widened and deepened the valley and created Ysgyryd Fawr's wedge shaped profile. One explanation for the large landslides is that they are the result of the oversteepening of these faces by moving tongues of ice which filled the valley and cut away the lower slopes of the adjacent hills, producing very steep valley sides. Once the buttressing effect of the ice was removed and the slopes thawed, gravity caused the slopes to become unstable and slip downwards, creating the scars and mounds you will see from the path. The name, Ysgyryd, means shattered or shivered. There are a number of problems with this interpretation and an alternative one is that the surfaces (bedding planes) between the successive layers of rock have been weathered and weakened over a very long period and eventually failed where the rocks were dipping gently south-west into the valley. This process may have been aided by the large volumes of melt water after 15,000 years ago.

The Monnow and Honddu Rivers no longer flow into the Usk. The Honddu was diverted by the Llanfihangel Moraine, which was formed by ice advancing northwards from the Usk Valley about 26,000 years ago, and now turns north to join the Monnow just north of Pandy. The Monnow then initially flows north-east along the line of the Neath Disturbance (see later), then south-east into the River Wye at Monmouth. The small Gavenny River which rises from springs at Blaengavenny, just south-west of Llanfihangel Crucorney, now flows south down the wide valley to Abergavenny. Much of the valley is floored by glacial till giving undulating, gently sloping ground which is often marshy.

Ysgyryd Fawr

Shortly after leaving the car park at the foot of Ysgyryd Fawr you enter the National Trust land at a gate and are walking on the St Maughans Formation of the Old Red Sandstone (ORS). You may be able to appreciate from the red/brown mud clinging to your boots that this Formation contains a lot of siltstone and mudstone. At the fork in the path (SO 32793 17015) look right, before you take the left branch, to see the mounds made by Yellow Meadow Ants on the south-east side of the hill, all

Figure 35: Intraformational conglomerate

Day 1. Ysgyryd Fawr (The Skirrid) to Llanthony

aligned to catch the morning sun. The ants are exploiting the soft sediments in the St Maughans Formation.

Where the path levels off (SO 32815 17064) there is a large, slipped boulder of laminated sandstone with a top surface of intraformational conglomerate. These conglomerates were formed when occasional powerful floodwaters scoured out channels through the sediments already laid down and ripped up pieces of mudstone and calcrete. As the flash flood subsided these pieces, together with sand, were deposited in the new channels. The mudstone pieces often erode out leaving irregular spaces in the rock.

At the meeting of many small tracks keep to the main path bearing half right and you will soon step onto the Senni Formation which here consists of thin beds of sandstone and thicker beds of intraformational conglomerate in which you can see pale chunks of calcrete. Most of the rocks here are red/brown but there are some green reduction spots which indicate where the red ferric oxide coating the sand grains has been reduced to green ferrous oxide when some of the oxygen atoms were grabbed by decaying plant or animal remains.

Figure 36: Rock face with calcrete base exposed by land slipping

At the point when you first see the path rising to the summit ridge, the marked path rises more gradually but to its left a stonier path climbs more directly to the ridge. If you take this left hand path and where it levels off move carefully left for a few paces, there is an interesting rock face exposed by land slipping. View it from the edge of the grass at SO 32792 17167 as the rough ground below the exposure is dangerously steep and loose. At the base of the face there is a thick, knobbly calcrete in which you can see the black tracery of fossilised Devonian plant roots. This is a fossil soil (palaeosol) which may have been exposed for thousands of years before it was inundated by the next flash flood which deposited the intraformational conglomerate above it. Between the two is a sharply defined boundary called an erosion surface.

Figure 37: Fossilised roots in calcrete palaeosol

The Land of the Beacons Way

Figure 38a: Cross-bedding along the summit path

Figure 38b: Cross-bedding near the summit

Above the conglomerate is thinly bedded sandstone, followed by more conglomerate and topped by more sandstone. In the conglomerate and sandstone you can see curved surfaces dipping to the south-west; these are called cross-bedding and are preserved cross sections through ripples migrating in the direction of the current, in river channels or lakes. They show that the currents here 400 million years ago were flowing from the north-east.

Figure 39: Exposure along the summit path showing erosion surfaces where younger river channels have cut across and down into older channels.

All along the path across the ridge there are exposures of thinly bedded and cross-bedded sandstones, intraformational conglomerates and cross-cutting channels. Everything we can see helps to build up a picture of braided, seasonal rivers, prone to flash flooding, pouring off high mountains to the north-east (the remnants of which are found in North Wales and Cumbria today) for tens of millions of years, depositing sediments on a low lying, coastal, semi-arid alluvial plain dotted with ephemeral lakes.

The view from Ysgyryd Fawr

The 360 degree view from the ridge and summit is spectacular. To the south across the Usk Valley are the Blorenge and Gilwern Hill, marking the northern rim of the

Day 1. Ysgyryd Fawr (The Skirrid) to Llanthony

South Wales Coalfield. The ORS is dipping gently south and forms the lower slopes of these hills. Their tops are made of Carboniferous rocks: limestones overlain by the Marros Group (formerly Millstone Grit) with the Coal Measures at the very top. To the west is Sugar Loaf, also originally part of the Black Mountains' plateau and also isolated by subsequent water and ice erosion. The upper half of Sugar Loaf is formed from the harder Brownstones Formation, with a few metres of tough Quartz Conglomerate on the summit. This layer protects the Brownstones and is responsible for the residual conical shape. The main Usk glacier cut into and steepened the south side of Sugar Loaf. The west and north sides were cut by a tongue of the glacier which branched off somewhere near Crickhowell and headed east-north-east exploiting strata already weakened and shattered by the Neath Disturbance. During the last glaciation Sugar Loaf was probably a nunatak, its top poking up through the surrounding ice. The dissected plateau of the Black Mountains, where you are heading, lies to the north-west.

North-east in the distance is the Golden Valley, a complex area of faults which is floored by soft sediments. On Ysgyryd Fawr itself the large landslip with its scars and debris mounds is usually well seen from the furthest part of the ridge.

Ysgyryd Fawr to Llanfihangel Crucornry

Figure 40: Rotational landslip north-west Ysgyryd Fawr

Figure 41: The landslip from the north

Once you start to descend you are soon back in the St Maughans Formation, but small old quarries working the Senni Formation can be seen above and to the left of the path. Around the east and north sides of Ysgyryd Fawr are many springs, marking where porous sandstone beds lie above impermeable mudstones, producing marshy areas. Further on the restored barn at Llanfihangel Court is built of Senni Formation sandstones which here produce quite thin, flat building stones.

Llanfihangel Crucorney is built on a glacial moraine which is the subject of ongoing research and is a key site for understanding the glacial history of the Black Mountains. The Llanfihangel Moraine is made up of two ridges and the larger, northern one is concave towards the north, which would suggest that it is the terminal moraine of a small glacier/ice-tongue flowing south from the mountains. But this appears not to be the case. The new research indicates that the northern ridge originally extended further northwards to fill the valley and is the terminal moraine of a branch of the Usk valley glacier which flowed northwards up the valley from Abergavenny. During a slightly warmer period the ice-front retreated southwards and a deep lake formed between the ice and the moraine. The climate then cooled again and the ice re-advanced over the lake bed to deposit the southern morainic ridge. When more sustained warming set in, probably around 15,000 years ago, melt water from thick snow cover in the Black Mountains poured south-east down the Honddu valley and eroded the northern edge

Day 1. Ysgyryd Fawr (The Skirrid) to Llanthony

of the Llanfihangel Moraine (which was almost certainly much larger) re-shaping it and giving it the appearance of being concave towards the north. The waters of the Afon Honddu were deflected north-east by the moraine, towards the Monnow.

Llanfihangel Crucorney to Hatterrall Hill

As you leave Llanfihangel Crucorney the bridge over the Afon Honddu is built of sandstone and is smothered in the Crabs Eye lichen (*Ochrolechia parella*), a species which prefers silica rich rocks.

After crossing the railway line and when you are walking up the field towards Great Llwygy Farm there are good views over the fence to your left of the eroded arcuate Llanfihangel moraine through which the railway runs in a cutting.

Behind the farm the hillside suddenly becomes steeper, this break of slope marks the transition from the softer St Maughans Formation to the harder Senni Formation. When you leave the wooded area and reach the crest of the open ground around SO 317 216 the Neath Disturbance lies almost at your feet, at right angles to the path, forming the valley running south-west behind Sugar Loaf and separating it from the prominent hill north-west of you on which is Twyn y Gaer, an Iron Age hill fort. This valley was created by earth movements along the Neath Disturbance and subsequent erosion of the shattered rocks.

Figure 42. The eroded arcuate Llanfihangel Moraine cut by the railway

41

The Land of the Beacons Way

North-west across the valley you can see the slipped hillsides above Cwmyoy. These landslides are also probably the result of long term weakening of the bedding planes, rather than glacial oversteepening of the hillsides.

After the cross roads at Trawellyd the path climbs up out of the valley onto the Black Mountains' plateau at Hatterrall Hill.

Hatterrall Hill

Figure 43: The valley following the Neath Disturbance, looking towards Sugar Loaf from above Great Llwygy Farm

Hatterrall Hill ridge is underlain by Senni Formation rocks which here are mainly sandstones with almost no conglomerates or cross-bedding. This suggests a quieter depositional environment than on Ysgyryd Fawr, possibly an ephemeral lake bed. The first of the Beacons Way artworks is set into a large slab of the Senni Formation whose lower part shows fine laminations, which also indicate a low energy environment with few animals living in and churning up the sediment. The thin beds made good flag stones and roofing tiles and were quarried, probably just for local use, in the many small workings along the ridge.

The Honddu valley below is a beautiful example of a U-shaped, glacially scoured valley. This scouring probably happened hundreds of thousands of years ago because many observations, including those at the Llanfihangel Moraine, suggest that the Black Mountains lay beyond the edge of the ice sheet in the last glaciation, 26,000 – 15,000 years ago. The more fertile fields of the valley floor are either on glacial or periglacial deposits or river alluvium. The path descending from the ridge towards Llanthony Priory runs across the lower part of yet another landslip, the edge marked by a line of springs which explain why the ground is quite wet in many places.

Llanthony

Llanthony Priory, at the end of this day's walk, is built of sandstone from the Late Silurian-age Raglan Mudstone Formation, the oldest of the ORS divisions. These rocks are not exposed here, apart from in the priory walls, but do outcrop not too far away, in the Wye and Golden Valleys. The stones are grey-brown, red-brown or purple and thicker than the local thin Senni Formation sandstones. There are some nice examples of intraformational conglomerate in the walls.

Day 1.Ysgyryd Fawr (The Skirrid) to Llanthony

Figure 44: Llanthony Priory built of sandstone from the Raglan Mudstone Formation

The Land of the Beacons Way

Day 2. Llanthony to Crickhowell

Maps: British Geological Survey 1:50,000 Sheets 214 (*Talgarth*) and 232 (*Abergavenny*)

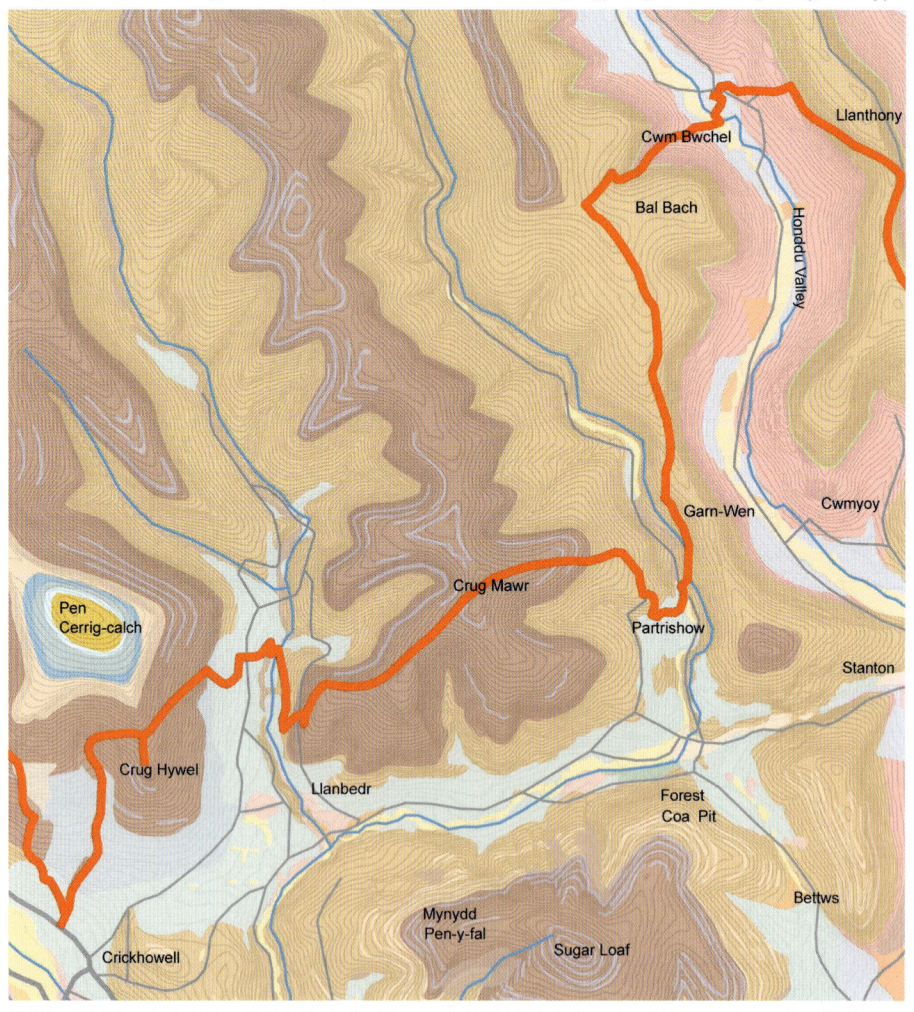

DiGMapGB50 geological data, British Geological Survey © NERC 2014. Contains Ordnance Survey data © Crown Copyright and database right 2014

Geology map for Day 2. with route marked

Day 2. Llanthony to Crickhowell

Highlights:
Cwm Bwchel calcrete
Cwmyoy and Darren landslides
Crug Hywel
View of Llangatwg escarpment

Synopsis of the walk
Once again the bedrock underfoot is ORS but today it does not all date from the Lower Devonian Period, there is some younger Upper ORS at Crug Hywel.

Shortly after leaving Llanthony, beside the path up Cwm Bwchel, there are blocks of calcrete which represent a Devonian fossil soil.

From the path over Garn Wen there are good views of the large landslides at Cwmyoy and the Darren.

Partrishow church is mainly built of ORS but the dressing stones are probably younger Twrch Sandstone.

From the top of Crug Mawr you can see across to Pen Cerrig-calch whose top is a remnant of Carboniferous rocks, left stranded on the Black Mountains by erosion.

Crug Hywel is a large block which has slipped down the southern flank of Pen Cerrig-calch. The rocks around the Iron Age fort on its summit are Upper ORS Quartz Conglomerate, 20-50 million years younger than the underlying Lower ORS.

South-west across the Usk Valley is Mynydd Llangatwg from whose quarries Carboniferous limestone was extracted for use as a flux in the many ironworks on the southern side of the mountain.

Details:
Llanthony
Shortly before the metal bridge over the Honddu River as you leave Llanthony, the path passes to the right of an old building which is being restored as Treats Café. The roof is made of Senni Formation sandstone tiles, about 4 cm thick, which are reputed to have come from Bal-Bach above. Stone tiles support many more lichens than do slates.

Cwm Bwchel

When the path crosses into access land, just after Cwm Bwchel Farm, you exchange the valley till for the St Maughans Formation underfoot. The path then climbs moderately steeply up Cwm Bwchel. As the stream gorge to your left starts to narrow there are a few different, dark blue-grey, knobbly rocks on the right hand edge of the path (SO 27882 27077). These are a calcrete, part of a thick limestone band called the Ffynnon Limestones which are found at the top of the St Maughans Formation over a wide area, even as far away as Shropshire. Like the other calcretes in the ORS they indicate a time gap of thousands of years during which the coastal plain stopped subsiding, no sediment was deposited and soils formed. In earlier times these calcretes (called bastard limestone) were an important source of agricultural lime to enrich fields far from conventional limestone deposits.

Figure 45: Calcrete block in Cwm Bwchel

After the calcrete blocks you are walking on Senni Formation rocks again, some thin bedded and others thicker and showing cross-bedding. The vegetation hereabouts is typical of ground on acidic, siliceous bed-rock: heathers, bilberry, crowberry, soft rush and tormentil, with cotton grass (a rush) in wetter areas.

Bal-Bach

From the summit ridge of Bal-Bach the higher top of Bal-Mawr is visible less than a kilometre away to the north-west. Its summit is underlain by the Brownstones Formation of the ORS and there is an obvious break of slope at the base, indicating a change in rock type.

Bal-Bach to Partrishow

Back in the Senni Formation, along the top between Bal-Bach and Garn-Wen are many small, old workings known as delves and marked as Pits (dis) on the OS map. From the stones lying around it appears that these yielded flaggy stones, probably for roofing or flooring.

Day 2. Llanthony to Crickhowell

Figure 46: The Darren- Cwmyoy landslip

From Garn-Wen southwards there are excellent and evolving views of the large and complex landslides which deformed Cwmyoy church and created the Darren (rocky edge) with a wide apron of disturbed ground below (see explanations in Day 1).

The upper part of the Cwmyoy slip is still moving significantly and the tilt of the church is still slowly increasing. There are more delves before the Stone of Vengeance, probably for local field walls.

Apart from its human story the Stone of Vengeance (aka the Bloodstone) has geological interest. It marks the interfluve between two very different valley forms, fuelling the debate around the extent and timing of glaciations in the Black Mountains. East is the open U-shaped Honddu Valley whereas the Grwyne Fawr Valley to the west is narrow and V-shaped, a style usually associated with water erosion rather than ice.

Just before Upper House Farm the bedrock exposed in the lane shows the gentle regional southerly dip of the strata. At Ty Mawr the house roof and the middle section of the barn roof are good examples of ORS roofing tiles, as is the roof of Tyn-y-Llwyn house at Partrishow.

Partrishow

Partrishow church seems to be built mainly of sandstones from the St Maughan's and Senni Formations. Most of the dressing stones however are different - paler, harder sandstone which contains rusty looking grains of an iron mineral, fragments of other rocks and black flecks of plant material. These stones probably come from a nearby outcrop of Twrch Sandstone (Marros Group/Millstone Grit) on Pen Cerrig-calch. They represent sediments eroded off the Wales-London-Brabrant High and deposited in braided river channels on a huge delta which extended southwards over much of South Wales.

Crug Mawr

As you climb up from Partrishow church towards Crug Mawr the start of the Brownstones Formation just about coincides with the wall below the open upland, where the slope steepens again. These Brownstones underlie the undulating but gently rising flat ridge stretching north-north-west from Crug Mawr to Pen y Gadair Fawr and Waun Fach.

When you reach the trig point on Crug Mawr a new panorama opens up. Westwards across the Grwyne Fechan valley is Pen Cerrig-calch whose upper part, as the name implies, is mainly Carboniferous limestone, the only exposure of this in the Black Mountains. Presumably Carboniferous limestone once covered the whole area from the Black Mountains southwards to the Coal Basin but the rest has been removed by erosion over 350 million years. Why a 45m thick remnant survived here is a mystery. The very top 20m of Pen Cerrig-calch is the Twrch Sandstone described above. Below the Carboniferous limestone and extending north-north-west as far as Pen Gloch-y-pibwr and Pen Allt-mawr is a 60m thick layer of the Upper ORS Quartz Conglomerate Group, forming a hard, flattish top to the ridge. This is at least 20 million years, and possibly as much as 50 million years, younger than the underlying Brownstones and separated from them by an unconformity representing

Figure 47: Benches along Pen Cerrig-calch

Day 2. Llanthony to Crickhowell

the time gap during which the Acadian Orogeny produced uplift and erosion. The top 10 metres of Crug Hwyel (the mound of Hywel), prominent to the southeast on the southern flank of Pen Cerrig-calch, is also Quartz Conglomerate and the presumption is that Crug Hywel is a land-slipped mass which became detached and slid down virtually intact over a vertical distance of 80m from the edge of Pen Cerrig-calch. From some directions this looks obvious and from others improbable. The timing of the slip is uncertain but probably sometime since the ending of the last glaciation 15,000 years ago. The strata are not badly disturbed so the movement may have happened gradually and possibly in stages. The slip plane is likely to have been a mudstone layer within the Brownstones which became saturated with water when the ice melted. You can see three mudstone layers within the Brownstones forming prominent benches along the side of Pen Cerrig-calch to the west.

South across the Neath Disturbance (responsible for the section of the Grwyne Fawr valley running roughly east-west) is Sugar Loaf, also capped and protected by the Quartz Conglomerate.

All the way down from Crug Mawr you are walking on the Brownstones until you reach the glacial deposits in the valley. The rising path to Crug Hywel also traverses these rocks.

Crug Hywel (Table Mountain)

As already noted, Crug Hywel's top is slipped Quartz Conglomerate, mainly visible as large blocks which dip at a slightly steeper angle than those on Pen Cerrig-calch above. You can see cross-bedding and ripple marks on some of the blocks. The ripples are fairly symmetrical which suggests a tidal

Figure 48: Cross-bedding in Quartz Conglomerate, Crug Hywel

Figure 49: Ripples on Quartz Conglomerate, Crug Hywel

or lake environment, possibly an estuary, rather than currents in a river. The Iron Age hill fort occupies the entire top and is surrounded by an impressive defensive ditch cut into the rock. Beneath a modern shelter near the entrance is the remains of an Iron Age hut and several levelled platforms inside the western ramparts mark the positions of more huts.

Crug Hywel to Crickhowell

From Crug Hywel there is a good view south-west, across the Usk Valley, to Mynydd Llangatwg (Llangattock) whose north facing escarpment is formed from thick beds of Carboniferous limestone. These 350 million year old limestones formed a huge coastal carbonate ramp dipping southwards in shallow water from the shoreline of ORS sediments towards a deep sea lying across Devon and Cornwall. This ramp contained a range of environments: brackish lagoon, inter-tidal flat, oolith sandbar, barrier island and clear shallow sea. Each of these environments produced a different type of limestone, all now exposed in the abandoned quarries facing you. The quarries first produced limestone for agricultural use and later as flux in the ironworks at Brynmawr, Nant y Glo and Ebbw Vale to the south of Mynydd Llangatwg. A network of tramways wound around the east side of the mountain, many still visible and walkable today, and inclined planes carried limestone and iron down to the canal. The younger, higher and larger quarries worked the limestone mainly as construction stone. Above the limestone on Mynydd Llangatwg are Marros Group rocks and above these, forming the top, is the Farewell Rock. This is a thick sandstone layer at the base of the Coal Measures which signified farewell to the promise of ironstone and coal to the miners.

At the western end of the escarpment are the spectacular cliffs of Craig y Cilau which tell a dramatic story. The first chapter is the undercutting of the mountain side by a combination of the Usk glacier and the Cwm Onneu valley glacier. Next, when the ice decayed, the pressure on the mountain side decreased, joints opened up in the rocks and a large slice of the mountain failed and slid downwards, probably as a single event. Finally, a small glacier formed in the basin left behind the debris. This ice then deepened the basin which today contains an unusual raised peat bog (Waen Ddu National Nature Reserve).

As you descend from Crug Hywel towards Crickhowell and look north-west up the Usk Valley you can clearly see that the Usk glacier must have passed either side of Myarth (the now wooded hill) but also that ice must have passed over it in colder periods to sculpt its smooth, streamlined profile. Myarth is composed of Senni Formation rocks with a small cap of Brownstones.

Day 3. Crickhowell to Llangynidr

Maps: British Geological Survey 1:50,000 Sheets 214 (*Talgarth*) and 232 (*Abergavenny*)

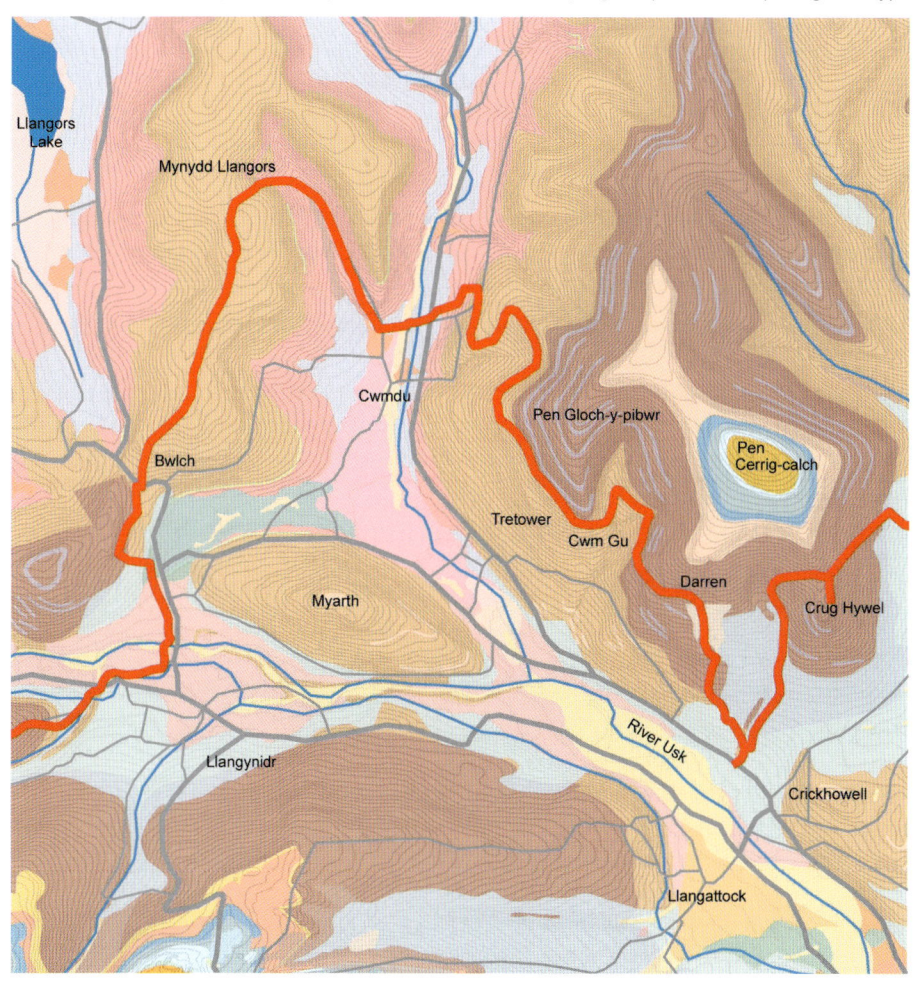

DiGMapGB50 geological data, British Geological Survey © NERC 2014. Contains Ordnance Survey data © Crown Copyright and database right 2014

Geology map for Day 3. with route marked

51

The Land of the Beacons Way

Highlights:
Darren
Cwm Gu
Llangors Lake

Synopsis of the walk

When you reach the open hill you cross the lower edge of another large landslip with cliffs above marking the back scar. This area is known as the Darren.

As the path descends towards Cwmdu the stepped profile of Pen gloch-y-pibwr is visible ahead. The stepped appearance is produced by four bands of mudstone which erode more readily than the harder sandstones.

The Rhiangoll valley at Cwmdu is floored by deposits left by rivers flowing out of a melting glacier.

From the grassy ridge of Cefn Moel, between Cwmdu and Bwlch you can see Langors Lake which is the remnant of a much larger lake formed when the Wye glacier to the north met the Usk glacier to the south and blocked the natural outlet of the lake.

From Bwlch you descend into the Usk Valley which was occupied by a large glacier during the last glaciation.

Details:
Crickhowell to Darren

Most of Crickhowell is built on the alluvial plain of the River Usk but as soon as you start walking up the lane by the Red Idigo Restaurant (formerly the White Hart), the Senni Formation is underfoot, although hidden here by a blanket of glacial till. Departing from geology, just beyond Pregge Mill you can see the remains of the leat bringing water to the mill from the stream on the east. At Gwerndale Farm the break of slope to

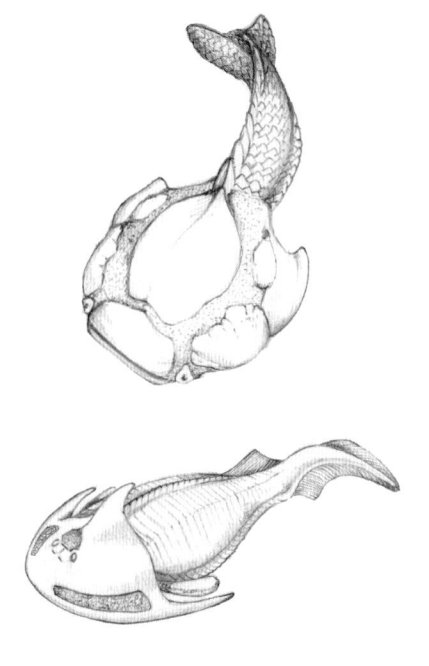

Figure 50: Representations of jawless fish

Day 3. Crickhowell to Llangynidr

the right of the track is doubly important: firstly it marks the transition to the harder and therefore steeper Brownstones Formation; secondly it is the line of the Neath Disturbance (see Day 1). The Brownstones to the east-north-east of the Disturbance have been thrown down about 200m relative to the Senni Formation to the west-south-west. In a small quarry just after the farm there are good exposures of Brownstones with cross-bedding indicating that they were deposited in fast flowing braided river channels. About 500m to the west is a disused private quarry in the Senni Formation where a rare example of a Lower Devonian freshwater jawless fossil fish with an armoured head-shield was found. The Devonian Period is also known as the 'Age of Fish' because it was during this time that fish diversified into many species, some ferocious predators and others with impressive defensive armour. The rivers and lakes of South Wales would have teemed with fish, but very few were preserved as fossils in that harsh, erosive environment.

To the south, across the Usk Valley lies Mynydd Llangatwg which was described in Day 2.

Beneath Darren

When you reach the open hill and contour to the north-west above the hill fence you will be crossing the bottom of an extensive area of uneven ground with tumbled rocks poking out of the bracken. This is yet another landslip which came from the south-west edge of Pen Cerrig-calch above. The Usk glacier sliced off the southern spur of the mountain leaving a steep face which subsequently became unstable. The timing of the slip is, as usual, uncertain, perhaps when the thaw came, perhaps only 5000-3000 years

Figure 51: Darren and the landslip

The Land of the Beacons Way

ago after weathering had widened the joints in the rocks. The disruption of the Neath Disturbance could have played a part in the instability too. Most of the slipped rocks are Brownstones with some Quartz Conglomerate. Some blocks have just tumbled down but in places there are big chunks of hillside which have slid down *en masse* on a slippery layer (usually mudstone) and rotated in a backwards facing curve as they did so. This area is also called Darren (rocky edge).

Darren to Cwmdu

As the path starts to descend, the stepped profile of Pen Gloch-y-pibwr appears ahead. Erosion of the four mudstone bands within the Brownstones has produced the risers, with the harder sandstones remaining as the flats of the steps. Next you enter Cwm Gu and cross a stream coming from a spring just above: water percolating through the Brownstones meets an impervious mudstone band and escapes to the surface. Cwm

Figure 52: Glacial Cwm Gu

Gu may have been excavated by a tongue of ice descending from the top of the Black Mountains during one of the early glacial episodes. As explained in Day 1 there is no evidence that the Black Mountains sustained an ice cap in the last main glaciation. There is however evidence, in the form of glacial erratics from the Builth Wells area found in the Rhiangoll valley, that Wye Valley ice spilled over the col at Pengenfford to the north and penetrated southwards into the Usk Valley.

If you stop to look at Cwmdu church, do notice the building stones as well as the inscribed stone. The main fabric is red brown/purple sandstones from the St Maughans Formation which underlies the superficial deposits on which Cwmdu is built. The dressing stones are different, they are the harder, well cemented pale grey sandstones which were probably brought from the Namurian layer on the summit of Pen Cerrig-calch, as were the dressings at Partrishow church seen on Day 2.

Where the Rhiangoll Valley widens south of Cwmdu there is a large spread of glacio-fluvial (rivers fed by a melting glacier) deposits forming flat, gently sloping ground which is seldom waterlogged and is good for agriculture.

Figure 53 Cross-bedding in Namurian sandstone Cwmdu church

Cwmdu to Bwlch

You soon climb back onto the St Maughans and then the Senni Formations as you walk up the lane towards Mynydd Llangors, turn south over Cefn Moel and descend to Bwlch. From this path you can see Llangors Lake to the north-north-west. This is no ordinary lake. Apart from being still the largest natural body of water in South Wales, it was once about six times larger and 30m deeper than at present. The water would have reached part way up the high ground you are now walking along, and as far as Bwlch in the south. During the last main ice advance the area that is now the Llangors basin was the setting for a series of confrontations between lobes of the Usk glacier advancing from the south-west and lobes of the Wye glacier from the north. The story is both complex and difficult to unravel but it now appears probable that the two ice streams met north of Llangors sometime between 26,000 and 20,000 years ago, at the height of the last glaciation, and their combined action over time scoured out the basin. Then, as temperatures rose slightly, both lobes retreated and melt water collected between them. The Usk glacier retreated fairly quickly back towards the Usk Valley but the Wye glacier retreated much more slowly, at only about 5m a year, and blocked the natural northwards drainage of the basin. Glacial Lake Llangors enlarged until the Usk ice had completely retreated into the valley and the lake was able to overspill southwards through two channels, one just west of Bwlch (not visible from our path). The lake stayed this size for about 600 years and would have been frozen over in the colder months but ice-free in the summer. Bwlch itself was probably covered by ice when the glacier was at its greatest extent.

Figure 54: Llangors Lake from Mynydd Llangors

Bwlch to Llangynidr

As you walk down the minor road and track from Bwlch into the Usk Valley you cross the north-east flank of Buckland Hill. This is a section of the Brownstones which has been dropped down towards the valley floor between two faults. When the track joins the road into Llangynidr the area of ground to the east is covered by hummocky glacial moraine deposits which were left when ice covered with debris melted irregularly. Llangynidr is likely to have been under at least 200m, and possibly as much as 400m, of ice when the Usk glacier was thickest. The glacier first deepened and scoured out the valley and then partially filled it with a thick layer of till sourced from the glacier itself and from melt-water streams. Since the glacier finally melted about 12,500 years ago the River Usk has cut down through the till so that, at Llangynidr bridge, the gently dipping layers of Senni Formation bedrock are again exposed in the river bed. The bridge which dates from about 1700 is constructed from the Senni Formation.

The power of a river to cut downwards depends on the sea level and the river's discharge. If the river stops cutting down because its power decreases the valley floor will stabilise at a new level. When down-cutting re-starts (either due to a fall in sea-level or to elevation of the land) the old valley floor will be left as a shelf on the valley side, called a river terrace. There are two such terraces visible at Llangynidr.

The prominent bare-topped hill visible in the west from the bridge is Tor y Foel which will be described in Day 4.

Figure 55: Glacial overflow channel at Bwlch

The Land of the Beacons Way

Day 4. Llangynidr to Storey Arms

Maps: British Geological Survey 1:50,000 Sheets 213 (*Brecon*), 231 (*Merthyr Tydfil*), 232 (*Abergavenny*) and *Fforest Fawr. Exploring the Landscape of a Global Geopark* (2014)

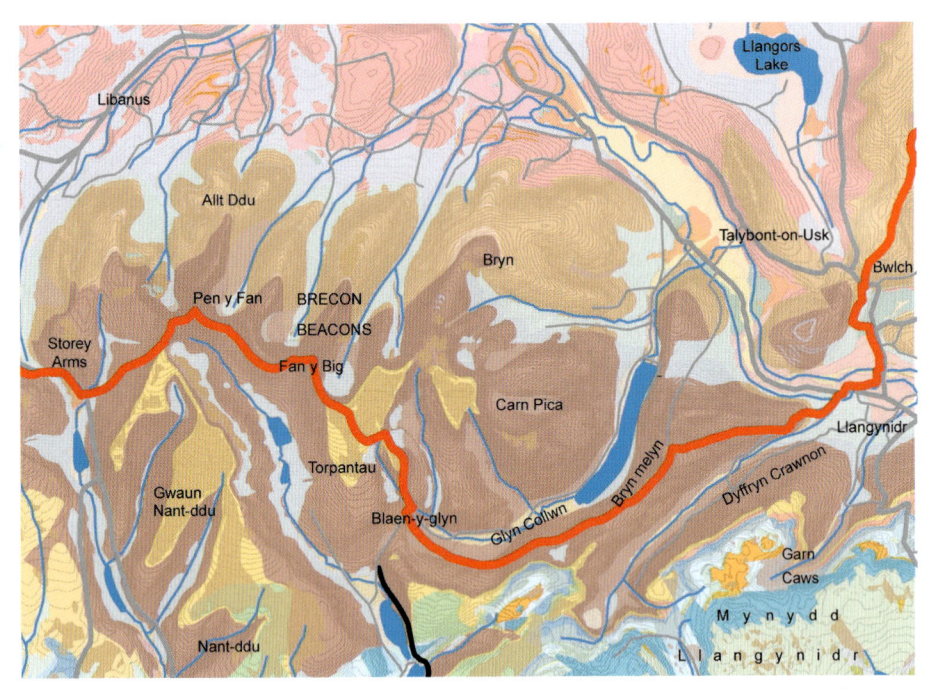

DiGMapGB50 geological data, British Geological Survey © NERC 2014. Contains Ordnance Survey data © Crown Copyright and database right 2014

Geology map for Day 4. with route marked

Day 4 Llangynidr to Storey Arms

Highlights:
Nant Tarthwynni
Steps up Craig y Fan Ddu
Brecon Beacons plateau and summits
View into Cwm Llwch

Synopsis of the walk
The day starts on glacial sand and gravel left by the Usk glacier but almost as soon as you leave the canal you move onto ORS, just north of the limestone forming the northern rim of the Coal Field Basin.

When you meet the lane from Talybont there are seats and a table made from fine blocks of ORS.

North-west there is a view up spectacular Nant Tarthwynni which was carved by a glacier flowing off the tops.

Glyn Collwn, which now holds the Talybont reservoir, is a U-shaped glacial valley.

From Blaen-y-glyn the path climbs steeply up to the central plateau of the Brecon Beacons. This path has been repaired with blocks from an older, muddier division of the ORS in which you can see the fossilized burrows of animals which lived in these sediments and also the cracks which formed when the mud surface was exposed to the fierce sun 400 million years ago.

The plateau was created by the interplay of uplift and erosion and then sculpted by ice sheets and glaciers which left a scalloped edge below which long, deep valleys radiate outwards.

On Pen y Fan's steep north face you see the layer cake appearance of the Brownstones Formation beds.

The summits of Pen y Fan and Corn Du are high and flattened because they are protected by a very hard layer of the Plateau Beds. They are also frost shattered as they were probably nunataks during the last glaciation.

Storey Arms is on the col where the last ice sheet divided to flow either north down Glyn Tarell towards Brecon to join the Usk glacier or south through Cwm Taf towards Merthyr Tydfil.

Details:
Llangynidr to Bryn Melyn

Today's route is described as starting from Lock 65 but you may have walked along the canal from the village centre first. The Brecon and Monmouthshire Canal is a triumph of engineering begun in 1796 to link Brecon with the Severn estuary at Newport. The 37 km (23 mile) section from Pontymoile Basin (near Pontypool) was constructed entirely along the 110 metre contour and does not need a single lock until it reaches Llangynidr. At Lock 65 the Afon Crawnon feeds the canal and then passes under it on an aqueduct to reach the Usk. The canal was an important link in the industrialisation of South Wales, transporting limestone, coal, iron and agricultural produce.

You cross the bridge and a stile into woodland. At the next stile the view opens up showing Tor y Foel to your right (west). Valley glaciers running north down Glyn Collwn and Dyffryn Crawnon to join the Usk glacier passed to west and east of this mountain. Ahead (south-south-west) is Mynydd Llangynidr whose lower slopes are in the Brownstones Formation above which are the hard Plateau Beds and Quartz Conglomerate Group, then Carboniferous limestone and sandstone and finally the highest points are formed of Twrch Sandstone. This is the northern rim of the South Wales Coal Field Basin.

At the gate by Llwyn-yr-eos the northwards view is up the ice carved Rhiangoll valley and that eastwards is along the Usk valley to Sugar Loaf with the Neath Disturbance passing to each side and fashioning its conical profile. Also to the east you can see Crug Hywell part way down the side of Pen Cerrig-calch.

When you leave the road to take the bridleway towards Bwlch-y-waun farm there is a good view up Dyffryn Crawnon which was probably shaped by a valley glacier originating from an ice-field over Mynydd Llangynidr and Mynydd Llangatwg. On the right at the head of the valley is Pen Rhiw Calch, the upper slopes of which, as the name suggests, are limestone. This is another piece of the Coal Field Basin rim which has been displaced northwards and downwards by the nearby Neath Disturbance and associated cross-faults. Bryn Melyn (yellow hill) is the hill on your left when you reach the lane from Talybont. Opposite the gate onto the lane are some fine blocks of ORS arranged as a table and seats. This ORS is a mellow pink colour and some of the seats have good examples of intraformational conglomerates showing where the ripped up chunks of softer mudstone have weathered out leaving depressions on the surfaces.

Figure 56: Nant Tarthwynni

Bryn Melyn to Blaen-y-glyn

From the seats and the first part of the road you now follow there should be a fantastic view west up Nant Tarthwynni to the cliffs of Craig y Fan with Carn Pica to the right and Allt Lwyd to the left. Nant Tarthwynni does not have a classic 'arm chair' cirque shape with a flat or scooped-out floor and a moraine across the outlet. Nevertheless it certainly would have channelled a glacier descending from the plateau above (Waun Rydd) in the last main glaciation and may have held a small one between 12,500 and 11,500 years ago.

Glyn Collwn, which now holds Talybont Reservoir is a beautiful glacial valley, in which the ice flowed north to augment the Usk glacier.

The signpost for Abercynafon is now placed a few metres before the bench mentioned in *The Beacons Way* but it is worth walking that bit extra to see the stone marking the Bryn Oer Tramroad. This is large block of Coal Measures sandstone containing pieces of coalified plant remains.

Figure 57: Plant fossils, some coalified, in the Bryn Oer Tramroad stone

The Land of the Beacons Way

Figure 58: Craig y Fan Ddu and Blaen Caerfanell

When you reach the Taff Trail the steep escarpment of Craig y Fan Ddu comes into view to the north-west and you see progressively more of the glacial cirque of Blaen Caerfanell as you walk. Craig y Fan Ddu is a southwards projecting spur of Brownstones Formation cut away on each side by ice and water but capped and protected by a hard layer of the Plateau Beds.

Blaen-y-glyn to Fan y Big

The steep, eroded path from Blaen-y-glyn up to Craig y Fan Ddu has been repaired with stone brought from Tredomen Quarry north of Llangors. The local rock type is the Brownstones Formation but the imported blocks, although still Lower ORS, are from the older and muddier St Maughans Formation. Tredomen Quarry is renowned for trace fossils (impressions left by living organisms rather than the actual organisms) and the path has many beautiful examples. There are burrows made by animals of various sizes burrowing through the sediment searching either for food or for damp places in which to lie low during prolonged dry periods. The animals were probably creatures resembling either large millipedes or small lobsters. These early Lower Devonian sediments were stuffed full of life and show that animals were successfully invading the land for the first time in the history of the Earth. Some blocks look like giant hot-cross buns, the crosses are desiccation cracks across muddy sediments where the cracks later filled with sand which now stands proud because the blocks have been placed upside down.

Day 4 Llangynidr to Storey Arms

Figure 59: Horizontal burrows

Figure 60 Infilled mud cracks in the undersurface of a mudstone bed

Figure 61: Horizontal burrows on the sole of a sandstone unit with the top of the underlying siltstone still adhering

The surfaces of the steps are eroding quite fast so they may look different to these images.

Beaconites barretti is the name given to the form of burrow, named after the Beacon Mountains in Antarctica, not the Brecon Beacons. The owner may have been an ancient lobster like animal.

Part way up the steep climb you could pause and divert 200 metres to the right to view the Craig y Fan Ddu Stone Rows at SO 05637 18058. There is a horizontal path to the right immediately after the upper fence corner. After 200m you cross a small stream and about 50m below you there is a group of smallish upright stones forming three roughly parallel rows with other random stones suggesting that the rows were

Figure 62: Vertical and horizontal Beaconites barretti burrows on the edge of a step, possibly made by the same animal

Figure 63: A large Beaconites barretti burrow, 17.5cm in diameter, backfilled by the animal as it tunnelled and appearing domed as it is seen from below.

Figure 64: The sunken tops of Beaconites burrows

Figure 65: Stream cutting through the moraine ridge in Blaen Caerfanell

originally longer. This monument is either from the late Neolithic or Early Bronze Age

At the top you are on the extensive plateau of the Brecon Beacons, created by the interplay of uplift and erosion and later deeply dissected by glaciers which carved out five spectacular trough head valleys to the north and northeast and three to the south. These valleys were formed during the main glaciations, either by fast moving ice streams within a covering ice sheet or by large valley glaciers. The moraine ridges within each main valley would have formed during the Younger Dryas Stadial. The path traverses the tongue of Plateau Beds rocks to reach the point where the Caerfanell stream cascades over the edge exposing ripple marks in its bed. Looking down from here into Blaen Caerfanell you can see an arcuate glacial moraine now cut by the stream. Looking north-east across the valley of Blaen y Glyn to the edge of Craig y Gigfran opposite you, there is an expanse of disturbed ground marking a landslip extending from the edge down to the stream.

You now cross the Gwaun Cerrig Llwdion plateau, where there are good exposures of the Plateau Beds, and rejoin the edge at the head of Cwm Oergwm which contains four ridges, two of which are certainly moraines. The other two could be pronivial ramparts (debris that slides down and collects at the foot of a steep snowbed).

The Beacons Way follows the edge around to Fan y Big which has lost its Plateau Beds cap and therefore has a less prominent and more frost-shattered summit. There used to be a Bronze Age round barrow at the top, now no longer visible. The views should be magnificent.

Fan y Big to Storey Arms

From Fan y Big the path descends to Bwlch ar y Fan at the head of the third northern trough head valley, Cwm Cynwyn, crossed by two moraine ridges, the lower, clearer one cut by the stream.

From the point where you cross the Gap Road over the bwlch you enter Fforest Fawr Geopark.

Cribyn has also lost its hard cap and has been shaped by erosion and ice into a tilted cone. Between Cribyn and Pen y Fan lies Cwm Sere which does not contain an obvious moraine, probably because there is no plateau to the west on which snow could accumulate so no cirque glacier formed in Cwm Sere.

Pen y Fan's steep north face clearly displays the layer cake appearance of the laterally extensive sandstones of the Brownstones Formation. These rocks represent extremely wide braided river channels. The summit of Pen y Fan has retained the flat Plateau Beds top, where many of the bedding surfaces display well preserved symmetrical ripples, evidence that these sediments were deposited in a shallow tidal environment. Near the tops of both Pen y Fan and Corn Du, approximately coinciding with the transition from the Brownstones Formation to the Plateau Beds, there is a change from ice-scoured and smoothed bedrock below to angular, shattered blocks on the summits. This transition is probably a glacial trimline marking the maximum height reached by the Brecon Beacons ice sheet during the last big glaciation. The summit rocks would have projected above the ice and were intensely frost shattered.

From the top there should be a view south-east down Blaen Taf Fechan, the largest of the southern trough head valleys. Along its western rim the escarpment shows three distinct segments: the nearest and farthest sections, Craig Gwaun Taf and Cefn Cul, are steep because the valley below them contained active glaciers during the Younger Dryas. No glacier formed in the central section as there was no snow collection area above it, only a narrow ridge, so there is no rocky, glacier-abraided scarp.

Look northwest from Pen y Fan into Cwm Llwch, the only valley to hold a glacial lake, dammed behind the obvious crescent shaped moraine. The ground to the east of this moraine shows hummocks and indistinct ridges which might represent remnants of a

The Land of the Beacons Way

cirque moraine from a previous glaciation.

The Beacons summits were significant places during the Bronze Age. On both Pen y Fan and Corn Du, the modern cairns overlie Bronze Age barrows (dated to around 4000 BP), both of which contained burial cists holding cremation fragments.

The wide track now leads downwards over the Brownstones towards Pont ar Daf and Storey Arms. They are situated on the col between Glyn Tarell to the north and Cwm Taf to the south. This col marks the centre and ice-divide of the last great ice sheet which started to decay about 15,000 years ago. From here ice flowed either north down the glacial trough of Glyn Tarell towards Brecon or south through the Cwm Taf trough towards Merthyr Tydfil. This upper section of Cwm Taf follows the line of the Merthyr Church Fault. At the footbridge over the Taf Fawr, just before the end of the day, you can look up the valley to see where a triangular section of the east facing slope has slipped downwards, probably due to undercutting by the stream.

Figure 66: Pen y Fan, Cribyn and Fan y Big

Day 5. Storey Arms to Craig-y-nos

Maps: British Geological Survey 1:50,000 Sheets 213 (*Brecon*), 231 (*Merthyr Tydfil*) and *Fforest Fawr. Exploring the Landscape of a Global Geopark* (2014)

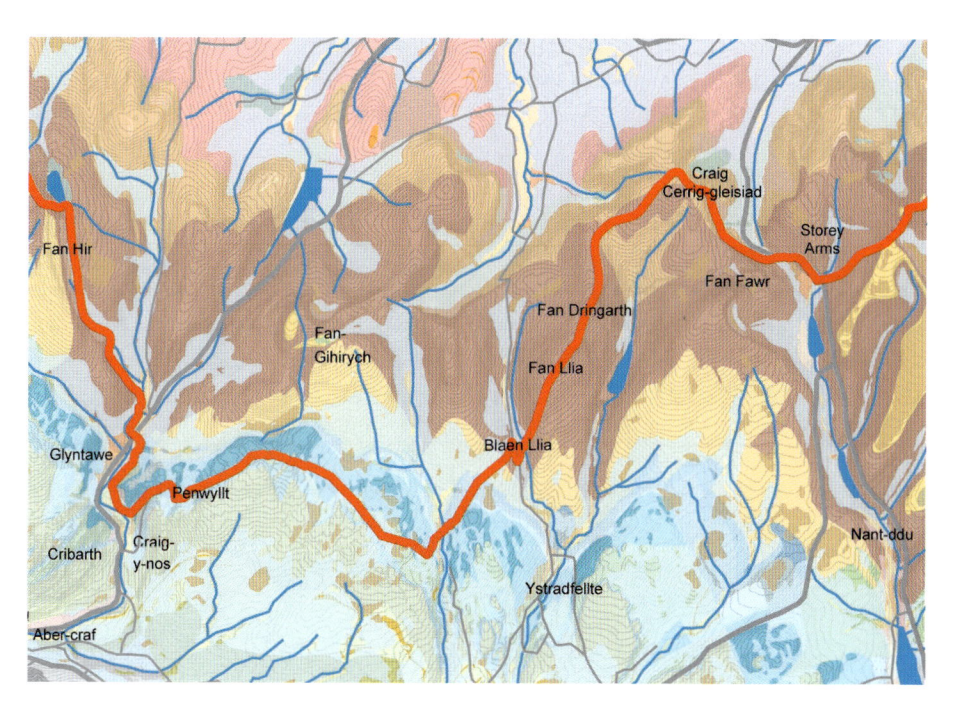

DiGMapGB50 geological data, British Geological Survey © NERC 2014. Contains Ordnance Survey data © Crown Copyright and database right 2014

Geology map for Day 5. with route marked

The Land of the Beacons Way

Highlights:
Craig Cerrig-gleisiad
Fan Dringarth landslip
Ogof Ffynnon Ddu
Cribarth and the Cribarth Disturbance
Penwyllt

Synopsis of the walk
The path starts by heading north along the western side of Glyn Tarell, the glacial trough down which the last ice sheet flowed to join the Usk glacier near Brecon.

For the first two-thirds of the day you are walking over the ORS Brownstones Formation. The highlight of this section is the view into the cirque of Craig Cerrig-gleisiad. The southern headwall is high and steep but the western headwall has collapsed, probably after the end of the Late Devensian glaciation, spreading debris to the north and east. Between 12,500 and 11,500 years ago a small cirque glacier formed in the area shaded from the sun, this left a lake which subsequently silted up and is now a peat bog. You can look down on the moraine ridge left by this glacier.

The path then heads south over Fan Dringarth where a large section of the eastern hillside has collapsed into Cwm Dringarth.

From Fan Llia there are views south to the Pennant Sandstone escarpment above Hirwaun.

You are then walking south down the dip slope of the strata so eventually, shortly before the standing stone Maen Madoc beside Sarn Helen, you cross onto the next youngest beds, the Carboniferous limestone.

After crossing the Nedd Fechan the path traverses the expanse of Pant Mawr where outcrops of limestone alternate with dolines and spreads of Twrch Sandstone. The reason for this mosaic of rock types is that large areas of the limestone have dissolved away underground allowing the overlying Twrch Sandstone to collapse into the voids.

Ogof Fynnon Ddu National Nature Reserve contains a rare example of limestone pavement where you can easily find fossils of large colonial corals, gastropods, brachiopods and sponges.

Underneath the reserve lies one of the deepest and longest cave systems in the British Isles. See www.ogof.net for a virtual tour.

There are good views of Cribarth, a much quarried anticline created by earth movements along the Cribarth Disturbance.

Day 5. Storey Arms to Craig-y-nos

Penwyllt is another interesting place. In the nineteenth century this was a thriving community where limestone and silica sand were quarried and furnace bricks manufactured in the now ruined brickworks.

Details:
Storey Arms to Craig Cerrig-gleisiad

The path rises steadily over the Brownstones Formation to cross above the waterfall in Nant y Gerdinen. Here a thick sandstone bed forms the top of the falls, the water has eroded the softer mudstones and siltstones below to create the fall. By the crossing the brick like pattern is caused by jointing. Above you thick sandstones higher up in the sequence form the scarps of Fan Fawr.

Figure 67: Channel and floodplain deposits

Along Craig y Fro at SN 970 206 there is a fine exposure of the Brownstones. A thick cross-bedded sandstone unit, representing a flash-flood channel deposit, is succeeded by a series of finer grained thin beds which were probably deposited either from extensive muddy floodwaters or in ephemeral lakes. The thick bed could have formed in hours whereas the thinner beds may have taken thousands of years to accumulate.

If you look down from the steep cliffs of Craig y Fro you will see that the A470 cuts through a definite, curved ridge enclosing a small peat bog and that there are further, less distinct ridges and disturbed ground extending north-eastwards down Glyn Tarrell. Glaciologists are still debating the origins of these features. One suggestion, backed up by a recent study of cores drilled at the site, is that they represent moraines which enclosed a small glacier during the Younger Dryas, between 12,500 and 11,500 years ago The low altitude of the site is against the formation of a glacier but there is a large plateau west of the headwall from where large amounts of snow could have been blown to feed a small glacier. Another suggestion is that the features are the result of a rotational slope failure landslide sometime after 15,000 years ago, at the end of the last major glaciation. There is evidence for and against this explanation too.

From the vicinity of the 629m spot height along the rim of Craig Cerrig-gleisiad there are extensive views to the west into the Senni Valley (with a disused quarry on the far side). This is the area after which the Senni Formation is named. You can also see the long ridge of the Plateau Beds on Fan Hir behind Fan Gyhirych. On the far side of Craig Cerrig-gleisiad cirque is Fan Frynach, between this and Fan Gyhirych runs the Cribarth

The Land of the Beacons Way

Disturbance, an important and long-lived zone of faulting and folding like the Neath Disturbance.

Craig Cerrig-gleisiad cirque

This is a spectacular and interesting place so it is a shame to cut the corner and head straight for Cefn Perfedd. Descend to the gate at SN 960 221, climb over the stile on the right and walk a few paces to the small, grassy promontory overlooking the cirque. The southern headwall crags rise to 622m and show deep gullies with largely inactive debris cones at their lower ends. These crags are cut by several faults which locally alter the dip of the strata, but the general dip is to the south-east, into the headwall. There are no equivalent northern crags so this cirque lacks the classic 'armchair' shape. The southern section of the western headwall consists of disrupted bedrock blocks which tilt backwards with tension cracks between the blocks. You are actually standing in the midst of these. There are also tension cracks visible on the grassy surface to the left of the path just before the stile. The northern part of the western headwall has a much gentler slope and a large triangular area of hummocky ground crossed by many approximately north-south ridges. On the floor of the cirque just beyond the base of this triangular area is a small depression containing a peat bog. East and north-east from the bog is an area of ridges

Figure 68: Craig Cerrig-gleisiad showing the steep southern headwall, the partially collapsed western headwall and some of the ridges on the cirque floor

and mounds with a large tongue-shaped lobe extending for more than a kilometre to the A470.

To cut a long discussion short, the following is the currently favoured explanation for the landforms:

Prior to the last major glaciation there was probably a shallower cwm in the area of the present cirque, possibly excavated in strata weakened by movement on the adjacent Cribarth Disturbance. The erosive action of the Late Devensian ice sheet and valley glacier excavated the cirque and oversteepened the walls. When this ice melted about 15,000 years ago the southern headwall remained fairly stable as the rocks dip into the headwall. The western headwall however was unstable as the rocks dip outwards so, once the support of the ice was gone and the ground thawed, it collapsed catastrophically spreading debris far to the east and north-east. Along the southern section of the headwall the bedrock blocks rotated backwards as they slipped, but the northern part may have slipped *en masse* down a bedding plane, rucking into ridges as it moved.

During the Younger Dryas Stadial a small glacier formed in the south-west part of the cirque, where it was shaded by the steep southern headwall and received sufficient snow blown off the Rhos Dringarth plateau above. These two factors can account for the development of a glacier at a relatively low altitude. This glacier modified the debris from the landslip and deposited moraine ridges.

After the last ice melted around 11,5000 years ago, when the mean temperature in Wales rose by about 8°C over only twenty years, there was probably a small lake impounded behind the moraine. The lake then slowly infilled to form the present peat bog.

The area below Craig Cerrig-gleisiad has been a favoured site for human occupation since at least the Bronze Age, and no wonder as it is sheltered by the cirque wall, the moraine ridges and landslipped terrain provided well drained soil and cover for hunters and there was a lake to attract animals. The cirque contains an Iron Age settlement, possibly on the site of an earlier Bronze Age one and the remains of a medieval longhouse as well as 19th century enclosures.

Craig Cerrig-gleisiad is a National Nature Reserve partly because of the arctic alpine plants which survive here, clinging to the ledges of the north facing cliffs. These plants are at the southern limit of their distribution in the British Isles, and only appear again, at altitude, in the Alps. Geology influences the plants in other ways; purple saxifrage, a lime-loving species, grows on the sandstone cliffs because the Senni Formation rocks here contain many calcretes. Peregrines, ring ouzels and ravens nest here.

Craig Cerrig-gleisiad to Fan Dringarth

The path heading towards Cefn Perfedd runs almost parallel to and 300m south of the cliffs of Craig Cwm-du. Nant Cwm-du possibly held another small Younger Dryas glacier, followed by a lake. You can look back (north-east) into this valley from the spot height on Cefn Perfedd. Crossing the top of Nant y Gaseg provides a good example of upland erosion by water, animals and people.

Just before the summit of Fan Dringarth you suddenly come across a very disturbed area of land to the left of the path extending down to the valley floor. This is another spectacular landslide, but little known as it is relatively inaccessible. The eastern side of the summit has broken away, probably due to instability when the glacier filling the Dringarth valley wasted away at the end of the Devensian glaciation. The slope failure is entirely within the Brownstones Formation, there are a few back-tilted blocks below the headwall but most of the material is just displaced downslope, little altered and now grassed over forming a great bulge over the Afon Dringarth.

There are good exposures of the Brownstones in the low backwall scarps at the top of the landslide. In one of these there is a nice example of soft sediment deformation,

Figure 69: Fan Dringarth landslip

probably formed when earthquake shocks caused an unconsolidated sandy layer to undergo liquefaction and then deform.

The upper valley of the Afon Dringarth contains many settlement sites dating from the 13th to the 17th centuries: enclosures, house platforms and the remains of longhouses. These could have been *hafotai* (summer dwellings). Purple moor grass may have been cut for 'mountain hay' as Gwair, the name of one of the streams means hay. There is also evidence for some older pre-historic sites.

Figure 70: Soft sediment deformation in the backwall of the landslip

Part of the valley was flooded when the Ystradfelle Reservoir was constructed between 1907 and 1914 to supply water for Neath.

Fan Dringarth to Blaen Llia

There are great views from the cairn on Fan Llia. The gritstone pavements above Ogof Ffynnon Ddu are visible to the west-south-west over the shoulder of Fan Nedd. The Pennant Sandstone escarpment above Hirwaun is prominent to the south. The Pennant Sandstone Group is a series of predominantly erosion-resistant sandstones deposited towards the end of the Carboniferous Period by meandering rivers flowing from high land to the south which had been pushed upwards by the Variscan Orogeny.

In the near distance to the south you can see the second stage of Alfred Russell Wallace's walk to Pen y Fan. Wallace was a self-taught great Victorian naturalist and the co-founder with Charles Darwin of the theory of evolution by natural selection. In June 1846 he was working as a land surveyor near Neath and decided to take his brother on a walk from Neath to the Beacons, to show him the countryside and to look for beetles. After leaving the Vale of Neath they crossed the Afon Dringarth at a now ruined farm situated just beyond the forestry, walked up the long ridge of Fan Fawr, down the other side to Pont ar Daf and then up Pen y Fan ('beetles very scarce'). At the footbridge the Afon Llia runs over the Brownstones Formation.

Blaen Llia to Ogof Ffynnon Ddu

As you turn left off the road onto Sarn Helen, Maen Llia a 4500 year old standing stone is visible on the northern skyline between Fan Nedd and Fan Llia.

You have been walking south down the dip slope of the strata and shortly before reaching Maen Madoc beside Sarn Helen you cross onto the next youngest beds, the Carboniferous limestone, but there is no immediate change in the landscape or the vegetation as the area is covered in glacial drift derived from the ORS to the north. You can however see limestone scarps across the hillside to the south, in an area called Carnau Gwynion where the ruins of 171 limekilns, most of them belonging to the commoners, are to be found. This was a busy area.

The path descends to the Nedd Fechan whose valley follows an almost north-south fault. On the west bank there are a series of natural limestone steps. The lower steps at the water's edge are of limey mudstone, deposited in a shallow carbonate-rich sea into which rivers poured mud from nearby land. The higher steps are of purer limestone laid down in a clearer shallow sea without muddy influxes. This limestone has dark grey fresh surfaces glittering with calcite crystals.

Figure 71: Limestone beds forming natural steps by the Nedd Fechan

Once you are on the path to Penwyllt there are limestone outcrops all around, many of which have been quarried on a small scale, and just before the ruins of Pant Mawr Farm there are fragments of limestone pavement. Interestingly the old house was mostly built of ORS although the boundary walls are of local limestone.

Shortly after the ruin shake holes/caprock dolines start to appear, which may seem strange because you should be walking over limestone, yet tumbled blocks of Twrch Sandstone are visible in the holes and the ground underfoot is normally wet and supports acidic vegetation. The reason is that all over this area the Twrch Sandstone, which should overlie the Carboniferous limestone (with an unconformity representing several million years of erosion between the two), has collapsed into extensive underground solution cavities in the limestone, probably in a series of solution-collapse cycles, dropping down a vertical distance of around 200m. Geologists call this process foundering. It is interesting that this whole area is called Pant Mawr (big hollow).

Pant Mawr was the site of a big rabbit farming venture in the early 19th century.

Figure 72: Lily of the Valley growing in the grikes of the limestone pavement

Ogof Fynnon Ddu (Cave of the Black Spring)

After you pass through the wall into Ogof Fynnon Ddu National Nature Reserve the bed rock soon becomes limestone again but the scarps forming the higher ground to your left (south) are of hard Twrch Sandstone. The top surfaces of these scarps have been scraped bare and polished by ice and can shine like mirrors in the sun.

The path passes by an area of limestone pavement which has been fenced to exclude grazing animals and protect the rare plants growing there. You can enter this through a stile by an information board and it is a good place to stop. As well as the plants there are spectacular fossils of the colonial coral *Siphonodrendron,* brachiopods (resembling a cross section through nested saucers), a gastropod (snail) named *Bellerophon* and *Chaetetes* a sponge .

Figure 73 Siphonodendron coral

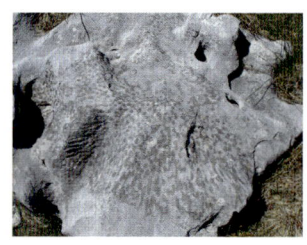

Figure 74: The gastropod Bellerophon

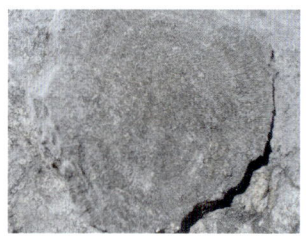

Figure 75: Chaetetes a fossilized sponge

Limestone pavements are not as common or as large in South Wales as they are in Cumbria and Yorkshire because the earth movements associated with the Variscan Orogeny fractured the rocks to a greater extent here. The conditions needed to form these pavements are still not fully understood. Thick beds of fairly pure limestone without too many joints are essential, but what happens then? It used to be thought that pavements formed when moving ice scraped away the overlying less pure and weaker beds leaving the top surface exposed to the elements, but this is now known to be only part of the story (see Part 1).

Beneath you lies the deepest (330m) and the third longest (50km surveyed) cave system in the British Isles. The discovery in 1946 revealed a new underground landscape of caverns, waterfalls, potholes, stalagmites, stalactites and more. This system was developed by solution, controlled by compression related fracturing, in a block of fairly hard/pure Carboniferous limestone. The great depth of the lowest passages is due to downcutting of the Tawe Valley by ice which lowered the water level and forced the tributary streams (above and below ground) to cut down also. The South Wales Caving Club has created an exciting virtual tour at www.ogof.net.

Ogof Ffynnon Ddu to Craig-y-nos

As you descend through the karst landscape towards Penwyllt there should be a good view ahead of the upfolded limestone ridge (anticline) of Cribarth. The folding must have happened after the limestone was laid down, probably about 300 million years ago during the Variscan Orogeny. Cribarth lies on the Cribarth Disturbance (mentioned at the start of this Day) and is part of the ancient Tawe Valley Disturbance along which the rocks have moved many times. The anticline actually continues north-eastwards across the valley to form the steep cliffs of Craig y Rhiwarth above Craig-y-nos but the Afon Tawe has cut a deep gorge through the fold, separating the two parts.

Day 5. Storey Arms to Craig-y-nos

Figure 76: The Cribarth anticline (behind is a syncline followed by another anticline)

Penwyllt

Towards the end of the descent towards Penwyllt, around SN 861157, you are walking down an inclined plane which was intended to transport coal and limestone northwards as part of the Brecon Forest Tramroad, the brain-child of John Christie the entrepreneur who purchased a large part of the Great Forest of Brecon from the government in 1819. Christie and his engineer soon realized that the steep gradient here made the scheme hopelessly uneconomic and re-routed the tramroad. Some stone sleeper blocks with drilled holes for the wrought iron pins which secured the rails remain by the path, but most were re-used.

Next the path crosses the line of the narrow gauge railway which brought silica sand and gravel from a pit 2.5 km to the north-east at Pwll Byfre to the silica brickworks at Penwyllt. The wagons were attached to an endless steel rope driven by a static engine at the brickworks. The brickworks operated from the 1870s to the late 1940s, producing fire bricks for furnace linings from crushed silica mixed with a small amount of limestone. The bricks not only went to the industrial areas of South Wales but were exported from Swansea to many parts of the world, including Egypt and Argentina. The works are now sadly ruined but the remains of the brick kilns, moulding and drying sheds and mixing pans can be seen alongside the railway 300m south of the caving club. See Figure 20 in Part 1.

The path continues to follow the line of the old Brecon Forest Tramroad. About 100 metres before the caving club buildings (former Powell Terrace) there is an example

Figure 77: Subsided tramroad with adjacent dolines

of recent solution subsidence. A short section of the tramroad has subsided, relatively undisturbed, while just to the left (south) of the path there is a line of developing dolines. The two must be connected and the subsidence here must be happening gradually.

There was limestone quarrying at Penwyllt before the Industrial Revolution. Commoners enjoyed the right to extract and burn limestone for domestic and agricultural purposes. When John Christie acquired land and mineral rights at Penwyllt he developed quarries and limekilns, intending to transport the lime north to Sennybridge on his new tramroad and use it to improve agriculture around the Usk Valley. The arrival of the Neath and Brecon Railway in 1866 meant that coal could be delivered in large quantities and limestone burnt on an industrial scale so new quarries and kilns were opened. These limekilns only shut down in the 1950s.

For over seventy years Penwyllt was a thriving community with a population of up to 500, the men employed in the brickworks, limestone and silica quarries, the limekilns and by the railway. Powell Terrace, the old Penwyllt Inn (the Stump), the platforms and waiting room, the general store cum post office (now a private house), decaying Patti Row and the dilapidated limekilns are all that remain.

If you missed the corals and brachiopods in the limestone pavement there are more in the big blocks alongside the quarry track from the car park.

The metalled road you then walk down was originally paid for by the opera diva Adelina Patti so that she could travel in comfort from her home at Craig-y-nos Castle to the station, where she had a private waiting room while awaiting her own private railway carriage.

When you turn off the road you are walking through Allt Rhongyr Nature Reserve, owned by Brecknock Wildlife Trust and managed for its special, unimproved limestone habitats. After leaving the reserve and shortly before you reach the lane there is a rock face to your left and at its base is the lower entrance to the Ogof Fynnon Ddu cave system.

Craig-y-nos is described at the start of day 6.

The Land of the Beacons Way

Day 6. Craig-y-nos to Llanddeusant

Maps: British Geological Survey 1:50,000 Sheets 213 (*Brecon*), 231 (*Merthyr Tydfil*) and *Fforest Fawr. Exploring the Landscape of a Global Geopark* (2014)

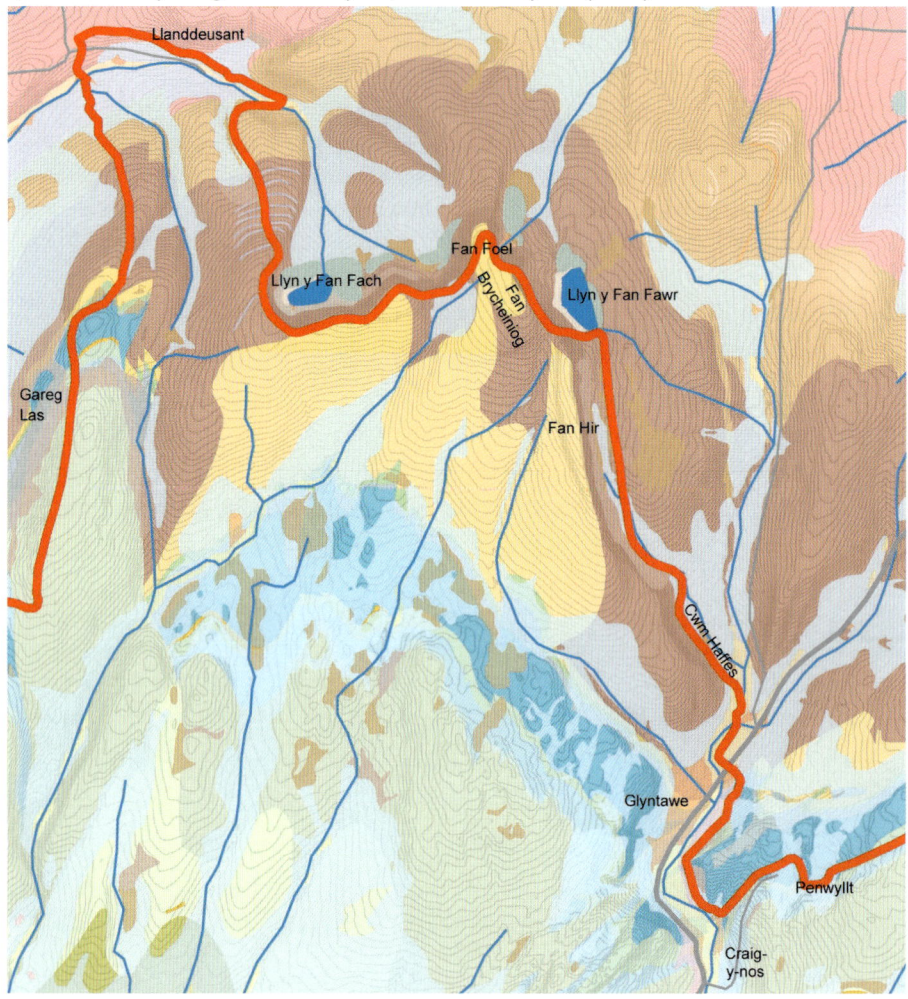

DiGMapGB50 geological data, British Geological Survey © NERC 2014. Contains Ordnance Survey data © Crown Copyright and database right 2014
Geology map for Day 6. with route marked

Day 6. Craig-y-nos to Llanddeusant

Highlights:
Craig-y-nos
Fan Hir moraine
Llyn y Fan Fawr
The escarpment from Fan Brycheiniog to Llyn y Fan Fach
Llyn y Fan Fach

Synopsis of the walk
Craig-y-nos lies in the core of the big anticline created by earth movements along the Cribarth Disturbance. This upfold extends from Cribarth across the Tawe Valley to Craig y Rhiwarth, but the river has cut a deep gorge through the fold, so providing a sheltered space for the castle and park.

Below the scarp of Fan Hir you walk along the top of a low ridge for over a kilometre. This is probably the moraine left by a long thin glacier which formed below the scarp.

Llyn y Fan Fawr probably occupies a hollow scooped out of the bedrock by a (relatively) fast flowing ice stream.

There are fantastic views from Fan Brycheiniog helping you appreciate the huge, many-faceted bulk of Y Mynydd Du.

The escarpment leading to Llyn y Fan Fach is a dramatic ice-sculptured feature.

Llyn y Fan Fach itself is a moraine dammed lake in a fine glacial cirque.

As you walk down towards Llanddeusant you enter the trough produced by the Carreg Cennen Disturbance.

Details:
Craig-y-nos
From the terrace at Craig-y-nos look north-east to the steep crag of Craig y Rhiwarth; near the top above the trees you can see that the limestone beds have been folded almost vertically by earth movements associated with the Cribarth Disturbance (part of the Tawe Valley Disturbance). The same folding produced the anticline of Cribarth to the south-west. The two structures were once continuous before ice and then the waters of the Afon Tawe cut a gorge between them. The Tawe Valley was deeply excavated by ice grinding down it along the line of rocks already weakened by the Tawe Valley Disturbance, so that, at the end of the last major glaciation, the rock floor was well below sea level. This meant that the upper Afon Tawe itself and streams draining

into it had to cut down rapidly to reach the new lower level of the main river, thereby creating gorges and waterfalls.

Just upstream of the footbridge over the Tawe the Nant Llynfell is seen joining the Tawe after draining about five square miles of fellside, sinking underground at Sinc y Giedd and emerging from the Dan-yr-Ogof cave system.

Craig-y-nos to Callwen

When you turn left through the kissing gate off the track to Pwllcoediog farm there is a deep valley visible in the hillside to the north-east. This is Cwm Haffes which marks the transition from ORS in the north to Carboniferous limestone succeeded by Twrch sandstone in the south. There is no fault here, the stream has eroded the softer beds of the Cwmyniscoy Mudstone Formation (formerly part of the the Lower Limestone Shale Group) at the base of the Carboniferous strata to create the cwm. The large sink hole by the path shows that the underlying rock is still limestone.

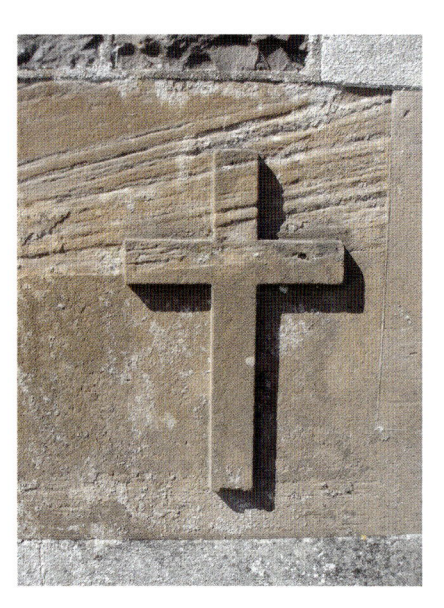

Figure 78: Cross-bedding at Callwen church

Ahead is Cefn Cûl, a mass of Brownstones isolated by erosion along strands of the Cribarth Disturbance passing either side of the hill. This was the site of another large nineteenth century rabbit farm.

On the corner of Callwen church next to the path there is an inserted block of ORS bearing a carved cross and a nice example of cross-bedding.

Callwen to Llyn y Fan Fawr

If you look back before the footbridge over the Tawe you can appreciate that Cribarth and Craig y Rhiwarth are indeed one fold sliced in two by the Tawe gorge.

About a kilometre after the waterfalls on the Nant Tawe Fechan you meet a fairly sharp crested ridge which extends north for 1.2km parallel to the scarp of Fan Hir. The southern end of this ridge curves and rises to meet the scarp and is cut through by the stream. This ridge, along which you will walk, is now generally interpreted as the moraine left by a narrow glacier which occupied the gully below Fan Hir in the Younger

Day 6. Craig-y-nos to Llanddeusant

Figure 79: The Fan Hir moraine cut by the stream. Above is the escarpment of Fan Hir topped by the Plateau Beds

Dryas Stadial. The moraine sits on a bedrock ridge which exaggerates its height. At its northern end the ridge peters out into low mounds.

East of the moraine there are many exposures of glacially smoothed Brownstones Formation bedrock.

Above to the west is the scarp of Fan Hir topped by the Plateau Beds with streams cascading over the hard edge.

Llyn y Fan Fawr to Fan Brycheiniog

Llyn y Fan Fawr is unlikely to be the site of a cirque glacier and moraine dammed lake as there is no convincing moraine and there is no protection to the south which a glacier there would have required. Instead the shallow hollow which now holds the lake was probably scooped out of the bedrock by a fast-flowing ice steam moving north to south in an early glacial episode.

Figure 80: Llyn y Fan Fawr

A word about the large leeches in Llyn y Fan Fawr, they are Horse Leeches *Haemopis sanguisuga* which cannot bite into mammalian skin but eat midge larvae and snails, so it is safe to bathe!

The lower end of the path up which you climb to Bwlch Giedd follows a fault which has thrown the rocks to the south down relative to those to the north. As a result the Plateau Beds have been lost for about 500m along the top, allowing erosion to carve the Bwlch into the softer Brownstones Formation.

Fan Brycheiniog

By the time you reach Fan Brycheiniog the Plateau Beds again cap the summit plateau as far as Fan Foel. Scattered over the path here are many quartz pebbles, some iron-stained pink or brown, which have weathered out of these beds.

On the gentle, western, upper slopes are good examples of patterned ground in the form of stone polygons.

The views from Fan Brycheiniog are extensive. Two kilometres to the north-east is Moel Feity with the source of the Tawe in the marshy ground between it and Llyn y Fan Fawr. The same distance to the north lies the source of the Usk beyond Fan Foel, with the Usk Reservoir in the distance. West is the escarpment of Picws Du and south the long edge of Fan Hir. The gentle dip of the land to the south and south-west follows the dip slope of the ORS beds which disappear underneath Carboniferous limestone in about 3 km. 16 km to the south-south-west is the prominent flat tilted top of Mynydd Allt-y-grug formed of Pennant Sandstone. Fan Brycheiniog is also a good position from which to appreciate the huge bulk and many facets of Y Mynydd Du.

Fan Brycheiniog to Fan Foel

After Twr y Fan Foel the scarp turns to face north-east and is scarred by two deeply eroded gullies which are actively cutting back into the edge. From the second of these gullies you look down at Gwal y Cadno (lair of the fox) where a small glacier was enclosed by an arcuate moraine on which a sheepfold has been built.

Figure 81: Moraine at Gwal y Cadno

Fan Foel

The flat, circular area with pieces of geo-textile protruding from a stone structure near the centre is protecting the remains of a Bronze Age round barrow. It was excavated in 2004 but was already damaged. The excavation revealed a central cist consisting of a stone box sealed by a capstone and containing the cremated remains of an adult, a young child and an infant; plus a pottery Food Vessel, a flint knife and meadowsweet flowerheads (identified by pollen analysis). This cremation was dated to around 4000 BP at which time the cist had been coved by a mound of turf, peat and soil. The barrow was used for a later cremation (around 3800 BP) of an adult and a juvenile, found together with fragments of a Collared Urn and a bone belt hook. The surrounding kerb of sandstone blocks probably dates from this second cremation. The position would have been a commanding one, visible from all the communities in the valleys below. After the excavation the site was backfilled with the intention of re-establishing a soil cover but weathering and erosion have re-exposed the area.

Fan Foel to Llyn y Fan Fach

The path descends the western flank of Fan Foel to Bwlch Blaen-Twrch, near the source of the Afon Twrch after whose valley the Twrch Sandstone is named. As at Bwlch Giedd a fault cutting the scarp has resulted in the loss of the Plateau Beds until Picws Du. The upper Twrch valley and the other valleys you can see to the south-west are fairly broad and shallow, suggesting that the last ice cover here was thin and probably cold-based (frozen to the bedrock therefore poorly erosive).

Below the splendid north facing scarp of Y Mynydd Du are seven depositional features presumed to date from the Younger Dryas Stadial. The exact origins of most of them are still being debated (see the section on Mynydd Du in *Classic Landforms of the Brecon Beacons*).

Llyn y Fan Fach

Figure 82: Llyn y Fan Fach showing the moraine around the stream exit. The ridge climbing the escarpment on the right may be another moraine or a pronivial rampart

The path along the scarp edge of Bannau Sir Gaer follows the edge of the Plateau Beds Formation outcrop until it approaches the south-west corner above Llyn y Fan Fach after which the Brownstones Formation is underfoot.

There is a fault at this corner, visible as an eroded gully. The lake occupies a beautiful cirque which would have accumulated a large amount of wind-blown snow from the big expanse of gently sloping ground above and to the west. There is no doubt that the cirque was occupied by a glacier whose moraine is visible beyond the northern shore, especially around the lake outlet. The lake is dammed but no longer used as a reservoir. From above you can see the line of the leat which collected extra water from the Afon Sychlwch to the north-east. There are spectacular debris cones cascading from gullies in the steep cirque walls.

Llyn y Fan Fach to Llanddeusant with Bronze Age cairns

Westward there is a good view of the limestone scarps of Carreg Y Ogof and the Twrch Sandstone ridge and Bronze Age cairns on Garreg Las further south (Day 7).

A series of mudstone bands within the Brownstones Formation crossing the path over Carnau Llwydion are manifest as wetter shelves in the terrain.

As you descend the marshy slope the Carreg Cennen Disturbance is running almost east-west along the valley ahead of you (see Day 7). To your right (east) is the U-shaped ice carved valley of the Afon Sychlwch leading towards the lake and, on your left, is the steep sided water cut gully carrying the Afon Garwnant.

When you reach Llanddeusant you have crossed the Carreg Cennen Disturbance and are back on the St Maughan's Formation.

Figure 83: Picws Du from Bwlch Blaen-Twrch

The Land of the Beacons Way

Day 7. Llanddeusant to Carreg Cennen

Maps: British Geological Survey 1:50,000 Sheets 212 (*Llandovery*), 230 (*Ammanford*) and *Fforest Fawr. Exploring the Landscape of a Global Geopark* (2014)

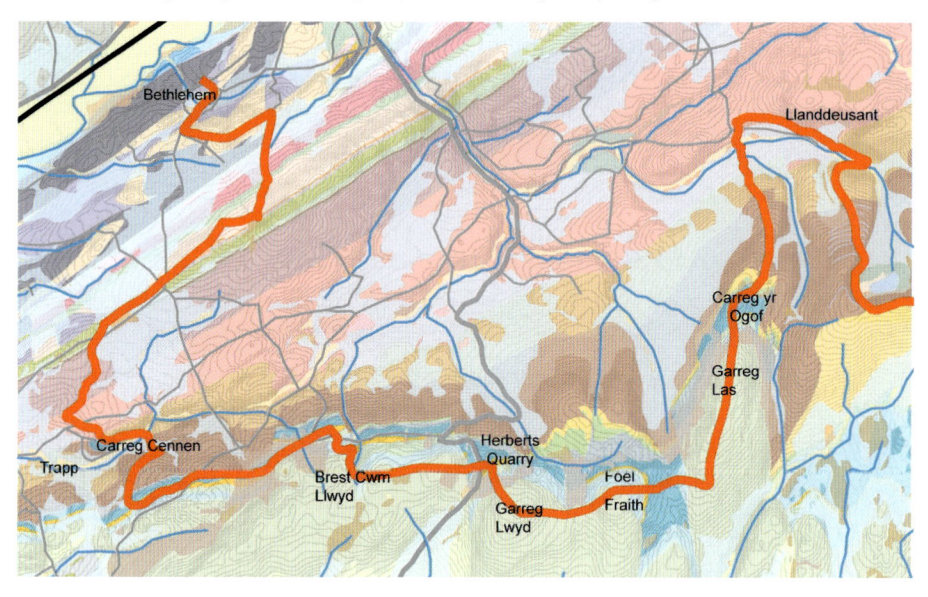

DiGMapGB50 geological data, British Geological Survey © NERC 2014. Contains Ordnance Survey data © Crown Copyright and database right 2014

Geology map for Day 7. with route marked

Day 7. Llanddeusant to Carreg Cennen

Highlights
Views
Carreg Yr Ogof
Herberts Quarry
Silica sand quarry
Carreg Cennen

Synopsis of the walk

This is a long, hard day with rough walking and some tricky navigation but it is worth considering the landscape from time to time as it is full of fascination and wild beauty.

Nowadays almost the entire area is remote, quiet and deserted, but in the 18th and 19th centuries it was intensely worked, busy, noisy and probably polluted. There were limestone quarries and kilns by every accessible outcrop, some as small scale enterprises and others as large commercial complexes. Quarry workers often lived nearby during the week in makeshift shelters. Silica sand was dug from large pits and transported to Brynhenllys Brickworks for making into refractory furnace bricks. Rottenstone was grubbed out of shallow pits by women and children who sold it for 1shilling per pony load to be used to polish the copper and tin smelted around Swansea.

The country was crisscrossed by tracks used by people, ponies with panniers and horse drawn wagons; and with inclines up which static steam engines hauled trucks. From 1779 the Llangadog Turnpike Trust maintained a route to the quarries on the northern edge of the mountain until the Brynamman to Llangadog turnpike (now the route of the A4069) was opened in 1819.

The earliest known use of lime in the area was for lime mortar used in building Carreg Cennen Castle in the 12th and 13th centuries. Small scale limestone extraction and burning with charcoal for agricultural use was widespread in South Wales from the 16th century. The industry expanded from the late 18th century as coal became available and transportation improved; lime was in demand for industrial processes as well as agriculture. Extraction and burning slowly declined in the early 20th century for many reasons. The complex of quarries now known as Herbert's Quarry (after the last known owner), which you will pass, only closed in the 1950s.

Old Red Sandstone, Carboniferous limestone and Twrch Sandstone were all quarried for building stone. Peat was dug at Cwar Penrhiw-wen, 800m SSW of Herbert's Quarry, in the 1940s and 1950s for use in the purification of domestic gas supplies. It is most likely that peat had been cut and used as a domestic fuel, and possibly for lime burning, around here for a long time. This area was of considerable economic importance for centuries because of its unique geology.

Details:
Llanddeusant to Carreg Yr Ogof

As you walk down the lane from the Youth Hostel to the Sawdde bridge you are entering the zone of the Carreg Cennen Disturbance, one of the very ancient and deep features cutting through the crust of South Wales. This fault zone is part of the Welsh Borderland Fault System which marked the south east margin of the Welsh Basin in Palaeozoic times. Movement, probably in many different directions over hundreds of millions of years, shattered the rocks along the fault making them very susceptible to ice and water erosion so creating a deep valley. The valley has been partially in-filled with glacial till but is still quite steep sided. The wooded valley of the Afon Mihartach, ahead and slightly to the left follows another fault, this one orientated north-south.

The stile to the open hill is a good place to pause. Looking north the valley following the Carreg Cennen Disturbance is immediately in front of you running just north of east to just south of west. To the west the Disturbance continues to Carreg Cennen Castle (which you can see with binoculars on its limestone crag) and beyond into Pembrokeshire. Eastwards the prominent cleft between the hills, Bryn Mawr and Waun Lwyd, marks the fault. In the further distance to the north-north-east, beyond the forestry around the Usk Reservoir, is Mynydd Myddfai. The rocks underlying the moorland of the summit ridge are Silurian sandstones, siltstones and mudstones, deposited along the shoreline of the Welsh Basin around 420 million years ago. The entire succession of Lower ORS formations is present between where you are standing and Mynydd Myddfai, spanning nearly 50 million years. South-south-east is the steep flank of Bannau Sir Gaer where the many mudstone bands within the Brownstones Formation show up as a succession of ledges. North-west is the Tywi valley extending south-west to north-east along a complex and wide zone of faults and steep folds known as the Tywi Lineament. This structure marked the boundary between the edge of the shallow shoreline shelf and the depths of the Welsh Basin.

All the Cambrian Mountains stretching to the north western and northern horizons are formed of sediments deposited in the Basin and subsequently forced upwards by earth movements. If you could have stood here about 450 million years ago you would have looked over a turbulent sea with huge underwater and emergent volcanoes in the distance (where Snowdonia is now) spewing out lava, ash, steam and pyroclastic flows.

Most of the rocks now scattered around are loose blocks but at SN 782 225 there is an *in situ* exposure of pebbly, cross-bedding Brownstones which dip moderately steeply to the south because of their nearness to the faulting (the general regional dip is about 5° south).

Day 7. Llanddeusant to Carreg Cennen

Approaching Carreg Yr Ogof it is worth leaving the path to look at the first of some small crags 80m to the left (SN 781 219). These are the Plateau Beds which you saw on Days 4 and 6 forming the erosion resistant summits of the Brecon Beacons and Y Mynydd Du. From here on they no longer form the summits because you are walking southwards; the strata are dipping southwards too and the planing action of erosion exposes younger rocks as you travel southwards. These outcrops are pebbly sandstones and conglomerates containing many iron-stained quartz pebbles. There are also a few darker red jasper pebbles which were eroded from deep marine sediments, most likely from around Anglesey. The well rounded pebbles in the conglomerates show that they were

Figure 84: Plateau Beds outcrop

Figure 85: Approaching Carreg Yr Ogof. Right to left: Devonian glacially smoothed Plateau Beds outcrops, low lying wet ground underlain by Carboniferous Cwmynyscoy Mudstone, hard Carboniferous limestones of Carreg Yr Ogof

91

transported quite a distance, and by water. The beds here dip to the north-north-west so have been distorted by the neighbouring faults (see below). A short distance west, in the vicinity of Carn y Gigfran (the raven's cairn), there is evidence that millstones were made from the pebbly sandstones. From now on the tops are of either Carboniferous limestone or Twrch Sandstone.

Then comes a break of slope with a flat, boggy area to the right of which are flat, glacially scoured expanses of the Plateau Beds. Cross-bedding within the rocks shows that the water currents flowed from the north, probably in a succession of braided streams on an alluvial fan at the edge of tidal flats. The reason for the marshy area is the nature of the layer above the Plateau Beds, the soft Cwmynyscoy Mudstone Formation (formerly part of the Lower Limestone Shale) which has been eroded into a hollow where peat could form because of poor drainage. The shales are the result of mud washed off the land deposited together with thin layers of calcium carbonate as sea levels began to rise at the very start of the Carboniferous Period.

From now on until you descend into the valley of the Cennen you will be walking along the northern rim of the South Wales Coal Basin.

Carreg Yr Ogof

A few paces later you are up among the hard Carboniferous limestones forming Carreg Yr Ogof (rock of the cave). There was extensive small scale quarrying, probably initially for agricultural lime but records imply that burnt lime for industry was exported south from here. The coffin route between Llanddeusant and Ystradgynlais (described in *The Beacons Way*) was very likely one of the routes used. No masonry limekilns are visible now but you can see the outlines of circular areas, tips of incompletely burnt limestone & pieces of coal which must have been transported from further south.

This is a good place for a break so there may be a chance to find fossils such as brachiopods (lamp shells), gastropods (snails) and a sponge called *Chaetetes* which lived on the shallow sea bed and often formed low reefs. There is also a bed of colonial corals as on Ogof Ffynnon Ddu (Day 6).

Figure 86: The sponge Chaetetes in a loose block

All these limestones were deposited in a large lagoon as the tropical sea slowly encroached on the Old Red Sandstone Continent. You may also find pieces of the succeeding layer, the Honeycombed Sandstone. This looks like it sounds, the spaces being where the limestone has dissolved away leaving a sandstone framework. A sea level fall which allowed rivers to carry sand from the land into the clear sea was responsible for this bed. The small cave within the younger, harder limestone near the summit leads into a passage extending for 580m.

Carreg Yr Ogof is completely surrounded by sandstone – Twrch sandstone to the south and Old Red Sandstone on the other sides. It is an isolated interloper, part of a block of crust which has dropped downwards at least 100m, probably in stages, between two faults. Geologists call this kind of feature a graben. Likewise Garreg Las (Twrch Sandstone) is the southern part of the graben, most of it is surrounded by either ORS or Carboniferous limestone. These faults are just two of a multitude of faults running north-south across the South Wales Coal Basin and adjacent areas. Why they have this direction is still open to debate; they may have originally been thrust faults due to oblique compression during the Variscan Orogeny and subsequently changed into extensional faults as the compression relaxed.

Figure 87: Inside the cave on Carreg Yr Ogof

South and south west of the trig point are several areas of limestone pavement. They are closely fractured and do not show the typical pattern of clints and grikes. In the low area to the east of the summit dissolution of the limestone along joints has produced shallow caves whose roofs have subsequently collapsed producing a karstic area of large dolines.

Carreg Yr Ogof to Garreg Las

It is geologically unclear why there is a valley between Carreg Yr Ogof and Garreg Las. The former is Carboniferous limestone and the latter Namurian Twrch Sandstone deposited in river channels on a vast delta. The boundary between these two major rock types is covered by glacial till and not exposed, but as you start up the north facing slope of Garreg Las the loose blocks and the series of low scarps are of Twrch Sandstone. You can find the fossilized impressions of small tree trunks or branches trapped in the channels, some forming log jams.

The Land of the Beacons Way

The long top of Garreg Las (blue rock) is strewn with frost shattered blocks of Twrch Sandstone, which have also been used to build the two large Bronze Age cairns. The cairns were originally solid structures but have been hollowed out as shelters for sheep and people. Many of the blocks show a peculiar fretted pattern of weathering. This may be due to subglacial weathering by water under high pressure. As the path starts to descend over Godre Garreg Las some areas of gritstone pavement show glacial striations (scratches), mainly orientated north-south (ice here in the last glaciation probably flowed southwards). This very hard rock cracks into regularly spaced joints when stressed by earth movements. These joints show clearly on the pavements and are nothing to do with ice.

Figure 88: Frost shattered Twrch Sandstone blocks on Garreg Las

Garreg Las to Garreg Lwyd

The low boggy ground between Garreg Las and Foel Fraith (speckled bare hill) is due to the continuation of the fault along the western side of Garreg Las which, as mentioned earlier, has dropped thick Twrch Sandstone down next to Carboniferous limestone with just a thin veneer of Twrch Sandstone. The strata have been fractured by the faulting, encouraging the limestone to dissolve below the surface and allowing the sandstone to collapse *en masse* into the voids, producing poorly draining ground and caprock dolines. Such occurrences are called foundered strata. This area is called Blaenllynfell and it was here in 1873 that the Black Mountain Company opened limestone and silica sand quarries using tramroads and inclines to reach Brynhenllys on the south side of Y Mynydd Du. The faulting and later foundering were probably responsible for the silica sand (see Pen Rhiw-wen). As you cross this wet area more limestone workings are visible ahead to the right on the north east flank of Foel Fraith.

Most of the ascent of Foel Fraith is on limestone where the grass is drier, less rank and without clumps of rush. The transition onto the Twrch Sandstone of the featureless top is marked by boggy ground, rough vegetation and dolines and these conditions continue over Garreg Lwyd (grey rock). Once again the col between the two and also the valley extending northwards follow fault lines. From here there should be a good view south-east across the valley of the Afon Twrch which runs north-north-east to south-south-west along another fault. This steep sided valley has been cut by the river which has also given its name to the Twrch Sandstone.

Garreg Lwyd to Pen Rhiw-wen

On the descent towards the road you are soon back on limestone and its easier vegetation. Although the lower land due west is Twrch Sandstone, the west flank of Garreg Lwyd is surrounded by limestone due to a major north-south fault, the Cwmllynfel Fault, which the A4069 (apart from the big bend) now follows. This fault, like the others, has probably moved many times, the net result being that all the rock layers to the west have dropped downwards about 100m relative to those on the east. Because these layers dip gently to the south it appears on the surface, and on the geological map, that the rocks also have moved laterally, but this is probably not so.

Depending on which path you follow down, you may see stone stripes on the hillside. These are a type of patterned ground, a periglacial feature.

Pen Rhiw-wen and Herbert's Quarry

When you reach the road you are on Carboniferous limestone with the substantial Herbert's Quarry area to your right. This is well worth an exploration and there is an interpretive trail around the site; as part of the CALCH project by Dyfed Archaeological Trust and the Brecon Beacons National Park. The area is possibly unique in South Wales for preserving a range of lime kilns spanning the transition from small pre-19th century clamp kilns through early-19th century intermittent use flare kilns to 20th century

Figure 89: View of Herbert's Quarry

masonry draw kilns. CALCH is aiming to unearth and record the history of the lime industry on Y Mynydd Du and has an excellent website at www.calch.org.uk.

About 30m east of the car park, near the tarmac track, is a small deposit of tufa forming in a stream draining from the lime kiln spoil tips. Tufa is crystalline calcium carbonate which can take many beautiful forms. At Herbert's Quarry tufa forms in a very unusual way. Water running through the limekiln spoil heaps dissolves the slaked lime to become a highly alkaline solution of calcium hydroxide (pH about 12). When this water runs out into the air the calcium hydroxide reacts with CO_2 to form calcium carbonate which then precipitates as it is insoluble in very alkaline solutions. There are many

Figure 90: Lime dram on rails inside a masonry draw kiln

Day 7. Llanddeusant to Carreg Cennen

Figure 91: Delicate tufa terraces forming in water draining from a lime kiln spoil heap. The golden brown colour is due to minute amounts of iron

intriguing tufa formations in the streams draining from the long northern edge of the quarries

Directly across the road (and the fault) are big silica sand pits in the Twrch Sandstone. It is thought that this hard rock was disaggregated into sand by weathering under tropical conditions sometime during the Tertiary Period (66-2.6 million years ago). This would imply that little glacial erosion occurred in this area throughout the entire Ice Age, otherwise the silica sand would have been scraped off. Low levels of glacial erosion mean that the ice was cold-based and frozen to the underlying ground. Cold based ice is due either to a very low ambient temperature, which is unlikely at this latitude, or to a thin ice cover so that the temperature at the base of the ice (which increases with depth) never reaches melting point. Other evidence suggests that the latter was the case over Y Mynydd Du. The Beacons Way skirts the edge of the workings but if you have time to go in you will see that much of the rock is now coarse sand containing many friable pieces and that the *in situ* unweathered blocks on the back wall dip steeply towards the fault. It may be that fracturing of the rock by the faulting facilitated the later deep weathering and disaggregation.

The detailed geology is even more complicated to follow for the rest of the day for several reasons. The strata forming the rim of the Coal Basin are cut by many north-south faults which displace the units either north or south, also expanses of foundered strata are common. As a result there are abrupt transitions from productive limestone crags with old workings to barren, wet sandstone areas.

Figure 92: Silica sand quarry

Pen Rhiw-ddu

The moorland is littered with blocks of Twrch Sandstone which have been frost-shattered, probably in the cold period after the ice cover had melted. The surface of some blocks display small dumbbell shaped depressions which may be the openings of the fossilised U-shaped burrows of a type of worm. There are also plant remain to be seen. Neither is that obvious so you have to be lucky and in the right place in the right light. The few areas of gritstone pavement are too weathered to show glacial striations.

Carn Pen-y-clogau

On the hillside east of the Bronze Age cairn many of the loose Twrch Sandstone blocks have formed stone polygons, another type of patterned ground.

South-west of Carn Pen-y-clogau there is an area of rottenstone workings at SN 71359

Figure 93: Periglacial stone polygons east of Carn Pen-y-clogau

1848 1; numerous small, shallow pits now mostly smothered by bilberry and mosses. You may pass through them depending on which faint track you follow. The OS map appears to mark Banc y Cerrig Pwdron as directly south of the cairn, but there are no workings to be seen in that area. Rottenstone started life as a shaley limestone lying immediately under the Twrch Sandstone. It is likely that, during the long process of sediment compaction and rock formation, strongly silicaceous fluids percolated downwards from the sandstone into the shale and set into a hard framework around the limestone fragments. The limestone then weathered away leaving a very light rock rather like a fine silica sponge which could easily be crushed to make a coarse, abrasive powder. There is no rottenstone visible on the surface today, it was collected by hand and too precious to waste.

Carn Pen y clogau to Brest Cwm Llwyd

The three Bronze Age cairns of Tair Carn Uchaf are visible ahead as you descend the hill. After walking west past the rottenstone workings you come to, and cross, the head of a valley (marking another fault) and turn north along a bridleway which initially follows the western rim of the valley. You are in foundered strata again and the small valley to the right contains more disaggregated sandstone and thin shale beds. The bridleway is the old way across Y Mynydd Du between Brynamman and Llangadog, known as the Bryn Road and, in places, is still well drained and cambered. To the east limestone scarps run along the hillside with associated workings and triangular waste tips. As the track bends to run west there is an area of foundered strata below covered in many small dolines. When you meet the road at Brest Cwm Llwyd there is a small quarry face where the limestone contains ooliths (but difficult to see). The rock face looks north making it an ideal habitat for maidenhair spleenwort and red algae, both of which flourish on shaded limestone. Just across the road there is another large area riddled with old lime workings, kilns and waste heaps.

Brest Cwm Llwyd to Carreg Cennen

When you leave the road the ground is foundered strata again. After 300m the track crosses a stream bed strewn with Twrch Sandstone blocks. This is the Afon Cennen flowing northwards down into the valley where it will turn through 90° to flow west along the Carreg Cennen Disturbance. To your left (south) the Cennen has cut a substantial gorge through the foundered Twrch sandstone. There is no fault here but when the river was swollen with postglacial melt-water it was powerful enough to cut through the hard Twrch Sandstone which had been weakened by collapse. The piles of rocks are a reminder of the stream's former power.

The path continues over foundered strata but there are limestone crags above to the south and shortly before you rejoin the road limestone reappears underfoot. Near the road there is a large banked settlement and enclosure complex called Bryniau. It appears to pre-date the nearby lime workings and is known to have still been occupied in the nineteenth century. Cultivation ridges can be seen running across the largest field. There is evidence of old workings all along the south side of the road until you turn north towards the castle. The highest ground to the south is the Tair Carn ridge which is an anticline of Twrch Sandstone. At each end of the ridge is a group of three Bronze Age cairns, they have been disturbed but not excavated and probably covered cremations. Hut circles and enclosures abound on these uplands. The area was probably more populated in the Bronze and the Middle Ages than today.

When you turn off along the path signposted to the castle you walk initially along a fault which has brought limestone on the right level with Old Red Sandstone on the left, but the limestone soon comes to an end and you are on the Brownstones to the north of the Coal Basin rim and heading down into the Carreg Cennen Disturbance to cross the main fault shortly before the footbridge over the Afon Cennen.

The first section of the path through Coed y Castell is on ORS which suits oak woodland. Just beyond the information board a gully crosses the path and heads down towards the river, this is one line of the fault which then follows the river west along the base of the limestone crag. As the information board tells you, the upper part of the woodland is on limestone and has very different vegetation, predominately ash.

Carreg Cennen

The castle was obviously sited here because the upstanding limestone crag with a steep valley to the south was an excellent position for observation and defence. This crag is an enigma; it ought not to be down here surrounded by ORS. It should be more than 500m higher than its present level. The inference is that it must have dropped this distance downwards between faults to form a graben in the middle of the ORS, probably during the Variscan Orogeny. But that is not the end of the story. Most of the surrounding ORS dips very steeply to the south-south-east but the crag limestone and a small area of Brownstones on its southern edge make up a lozenge shaped faulted block which dips more gently to the north-north-west and is bounded on both sides by strands of the Carreg Cennen Disturbance which has split to enclose it (where the gully crossed the path in the wood). In addition, the Senni Formation beds to the south of this fault-bound block are folded into an anticline. One interpretation is that, sometime after the graben forming event, this piece of country was detached from its original position on the rim of the Coal Basin and thrust in a north-north-west direction over the *in situ* ORS strata. These displaced beds originally formed a big

Day 7. Llanddeusant to Carreg Cennen

Figure 94: Carreg Cennen Castle and crag from the south-east showing the limestone beds dipping away from you to the north-north-west

anticlinal fold which was cut by the fault splaying southwards from the main thrust. Erosion then removed most of the folded rocks, leaving the harder limestone crag as a remnant of the northern limb of the anticline.

When you go through the gate by the castle entrance look left towards the castle to see a prominent hollow running east-west. This marks the position of the main thrust-fault, to your right is ORS. As you approach the Tea Rooms the rocks in the path are the very red, almost vertically dipping, thin beds of silty mudstone from the St Maughan's Formation. This vertical dip is in sharp contrast to the gentle southerly dip of the ORS elsewhere. You are now on the southern edge of the wide folded and faulted zone called the Tywi Lineament. This probably originated as a zone of extensional (pull apart) faults when the Welsh Basin was being formed by stretching of the crust on the north-west margin of East Avalonia in the late Pre-Cambrian to Early Cambrian Periods. During the collision of the Acadian Orogeny in the Early-Middle Devonian Period the

The Land of the Beacons Way

area was compressed so intensely that the contents of the Welsh Basin were forced upwards. Strata along the south-east margin were stood on end and in some places were actually pushed past the vertical and are now overturned and dip steeply to the north-west. This southerly limb of the Tywi Lineament is known as the Myddfai Steep Belt. The Beacons Way runs right across it but, after 400 million years of erosion, the topography is no longer steep. The Tea Rooms are built of Senni Formation blocks with Carboniferous limestone for the dressing stones.

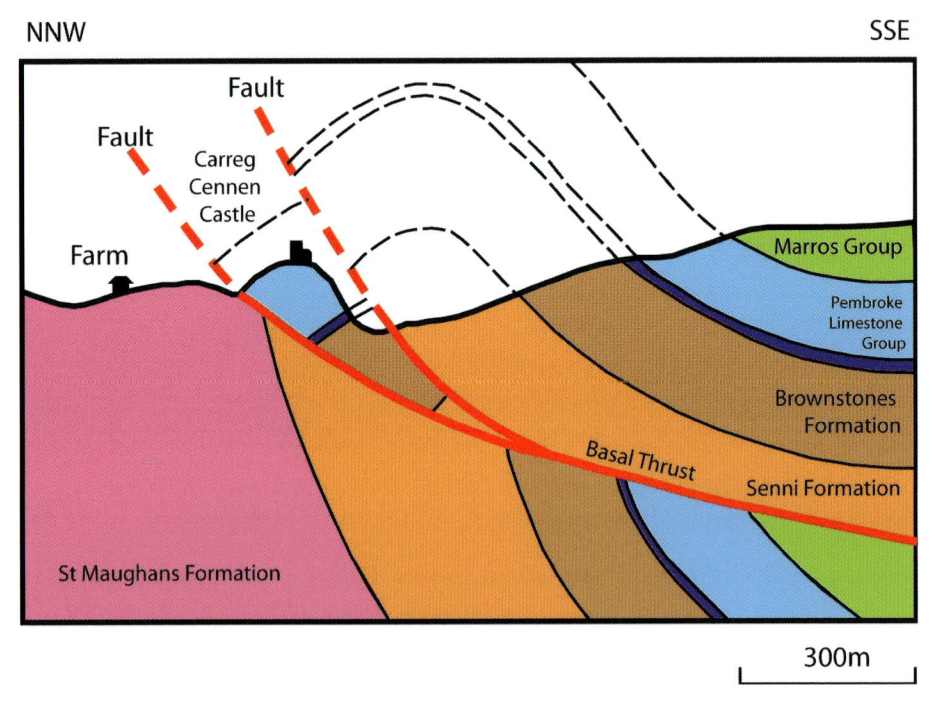

Figure 95: Cross section through the geology at Carreg Cennen Castle

Day 8. Carreg Cennen to Bethlehem

Maps: British Geological Survey 1:50,000 sheets 212 (*Llandovery*), 230 (*Ammanford*) and *Fforest Fawr. Exploring the Landscape of a Global Geopark* (2014)

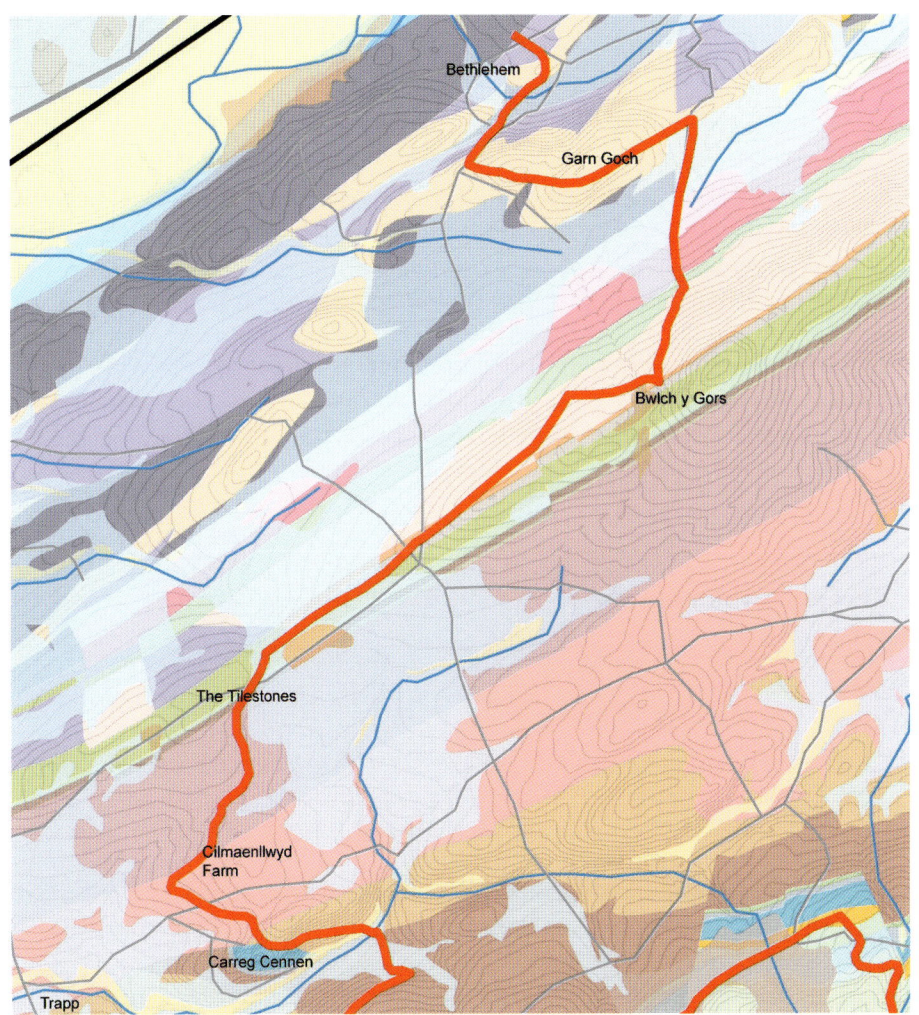

DiGMapGB50 geological data, British Geological Survey © NERC 2014. Contains Ordnance Survey data © Crown Copyright and database right 2014

Geology map for Day 8. with route marked

The Land of the Beacons Way

Highlights:
Carreg Cennen
The Tilestones
Garn Goch

Synopsis of the walk
Today you will be walking almost straight back in time through about 115 million years, from the Carboniferous Period into the Ordovician Period, from a shallow tropical sea, through a continental desert and into a deepening intracontinental basin sea.

Carreg Cennen Castle Carboniferous limestone crag is a geological oddity; see the end of Day 7.

The route then crosses the two oldest divisions of the ORS to enter the marine sandstones of the Silurian Period. This section may seem geologically boring but it does contain such gems as the Tilestones which form the crest of a ridge about 1.6 km after leaving the castle. They were an important source of roofing stone until the 19th century.

The next landmark is the dramatic hill fort complex on Garn Goch, built on and of the oldest rocks along the Beacons Way, the Ordovician Ffairfach Grits.

Details:
Carreg Cennen Castle
See the description at the end of Day 7.

Carreg Cennen to the Tilestones
As you walk up the lane from the car park to Castle View House the view to your left (south) shows the dipping Carboniferous limestone under the castle (the end of the graben is obscured by trees). The crag is surrounded by low lying Old Red Sandstone. The east-west valley along which the Carreg Cennen Disturbance runs is floored by the soft sediments of the St Maughan's Formation. Southwards the hill tops and ridges get younger, being formed from the Brownstones Formation, Carboniferous limestone and Twrch Sandstone, in that order.

On the left side of the track from Castle View House to Cilmaenllwyd Farm there are pronounced gullies in the fields. These represent fault splays from the disturbance.

Cilmaenllwyd Farm

The name Cilmaenllwyd means the hollow of the grey rock, but all the surrounding rocks look red. You have now reached the Raglan Mudstone Formation, the oldest division of the ORS. Underfoot in the farmyard are red outcrops of mudstone, coarse siltstone and fine sandstone beds which strike north-east to south-west and dip vertically. Some of the muddy beds contain grey specks and blobs. These are fragments of calcrete so perhaps there was once a larger exposure of grey calcrete, now quarried away or obscured, which gave the farm its name. The old farm buildings seem to have been built of the rock layer immediately underlying the Raglan Mudstone Formation, which you will meet shortly.

Figure 96: Flecks of creamy calcrete in thin, vertically dipping beds of red Raglan Mudstone

Just after the farm there are a number of small rock exposures in the left bank of the track where the beds dip steeply to the north-west. They are overturned. These exposures contain thin siltstone and sandstone beds which would have been deposited in shallow, ephemeral river channels on a coastal alluvial floodplain.

Very shortly the track makes a small dog-leg to the left. This is a good place to stop and look around. You are standing on a slightly elevated, elongated, lens shaped piece of ground underlain by the same river channel deposits. On either side of you the ground falls away slightly and is wetter, indicating softer, muddier beds deposited during large over-bank flood events when mud particles had a very long time in which to settle before the next event. The same cyclical alternation of hard and soft beds is manifested as ledges on the rising ground to your north. The boundary between the Raglan Mudstone and the younger, less muddy St Maughan's Formations is probably at the lower fence in the field to the south-east where there is a break of slope. To the north the Raglan Mudstones end where the slope steepens just below the rocky outcrops. The entire thickness of the oldest division of the ORS now extends for about 750 metres across your route because the originally flat-lying beds are tilted vertically. These 750 metres of ground represent around one million years of slow sediment accumulation.

To the north lies the rocky ridge of the Tilestones Formation. Soon you cross the stile by the stream and then marshy fields on the Raglan Mudstone Formation. Suddenly, for the first time since leaving the castle, there are stone walls instead of hedges and fences. Mudstones do not make durable walls. The ground is rising and there are piles of reddish sandstone rocks containing coarse layers of grit. These are the rocks used to build Cilmaenllwyd Farm. From the castle to this point you have walked back in time about 70 million years, to the edge of the Welsh Basin. These rocks (the Pont ar Lechau Formation) were the sediments deposited on the edge of a wetland which was periodically flooded by the sea.

The Tilestones

Although you are now going uphill you are actually walking down towards the sea, from younger muddy layers of rock, formed on land at the edges of huge alluvial fans, into older rocks laid down in the sea. The latter are more resistant sandstones so have survived as higher ground. Along the top of the low ridge to your right is the Tilestones Formation, seen in the distance earlier. This represent the edge of a delta protruding into the Welsh Basin, together with accumulations of sediment known as mouth bars which were deposited at the mouths of the river channels when fresh and sea waters met. The Tilestones define the transition from the older basinal marine rocks of the Silurian Period into the non-marine rocks of the Old Red Sandstone. They are flaggy, yellowish sandstones forming beds only 1-2 cm thick, often showing ripple marks. They also contain many crystals of sparkling mica which probably came from the erosion of Precambrian Avalonian rocks to the south-west. Crystals of mica are flat so thinly bedded rocks containing lots of them usually split easily. The Tilestones were formerly prized for roofing, locally and nationally. They roofed the great Cistercian abbeys of Strata Florida in Ceredigion and Abbey Cwmhir near Llandrindod Wells, both built in the 12th century. Once the railway system developed, Welsh slate which was easier to handle, replaced the Tilestones for roofing, but not until the 19th century. The Formation extends for about 50 kilometres, from south-west of Llandeilo to

Figure 97: Tilestone working

near Builth Wells, but is never more than 40 metres thick and is now tilted vertically in the Myddfai Steep Belt so the shallow pits from which it was quarried form a long, narrow feature across the landscape. These pits are only 3-4 metres in depth because near-surface weathering had conveniently loosened the cement holding the beds together. The deeper parts were too difficult to work. You can see such a pit a few metres to the left as the track reaches open ground and turns north-north-east to cross the line of the Tilestones.

The Tilestones to Bwlch y Gors

The Beacons Way now follows the boundary between the Tilestones to your right and the older Silurian fully marine rocks on the slight rise to your left (Cae'r mynach Formation). If you walk up to these Silurian outcrops and look back (west) you can see the upstanding

Figure 98: Bioturbation in the Cae'r mynach Formation

low ridge formed by the Tilestones being disrupted and offset in many places by the north-north-west to south-south-east trending faults which cut across it. The Cae'r mynach Formation is a shallow sea deposit in which the thin layers have been churned up by marine burrowing animals, leaving a lumpy appearance (bioturbation).

In the opposite direction (east) is the top of Pen y Bicws, the highest point on the Trichrug ridge (you can take a diversion there). This ridge is formed from the hard, gritty sandstones of the Trichrug Formation. Although these rocks are older than the Tilestones they too are continental deposits, probably the toes of alluvial fans pouring debris into a shallow wetland during an episode of sea level fall in the late Silurian Period.

Between you and Pen y Bicws is a broad valley marking another fault which has crushed the underlying rocks and encouraged erosion. These rocks are Silurian silty mudstones (Hafod Fawr Formation) laid down in deeper water before the sea level fall which gave rise to the Trichrug Formation.

The valley is now floored with clay rich glacial till, hence the marshy ground you must cross to reach the stile onto the minor road. Alongside the road are peat deposits which have collected in the fault hollow.

As you walk east along the road you may notice that the surface rises every few hundred metres. These rises signify splays of the main fault which drop the ground level down to the west and create low west facing fault scarps. As the road rises out of the valley, holly trees, which prefer well drained acidic soil, appear in the hedges. To your left (north) the downward sloping ground is also glacial till overlying the silty mudstones.

To your right (south) the line of the Carreg Cennen Disturbance is quite clear, passing

The Land of the Beacons Way

both sides of the castle crag and through the gaps between the hills to the east and west.

After the cross roads at SN 679218 the path continues to the north-east over the same mudstones, but on your right (south) the ground rises sharply to a fine wall built along the rocky crest of the hill (Carn Powell). The rocks forming the steeper crest are quite thick marine sandstones (the Gwar Glas Member of the Hafod Fawr Formation) which contain evidence of deposition on a shallow, storm affected shelf on the Welsh Basin margin. These rocks show that the basin was shallowing and filling with coarser sediments from the rising mountains. Another thing you may notice if you walk up to these rocks is that they are dipping less steeply, about 70° to the south-south-east, rather than vertically, indicating that you are approaching the flat top of the Myddfai Steep Belt (see the end of Day 7). The main wall is constructed from these moderately thickly bedded, greenish grey or purplish sandstones, but the cover band (horizontally laid thin stones below the coping layer) are Tilestones.

Bwlch y Gors to Garn Goch

At Bwlch y Gors the peaty area to the right of the low embankment marks another north-north-east-south-south-west fault which has caused the break in the ridge here. From the bwlch there is a signposted path up Pen y Bicws, almost to the top of the Trichug ridge (see above) where the Trichug Formation dips south-east at 60°. If you take this path you will see, a short distance to your left (north), a line of workings exploiting a bed of good quality sandstone with an upstanding outcrop of more pebbly rock behind them. From the top, twelve large pre-historic cairns are visible (two on Garreg Las, Carn Pen Rhiw-ddu, Carn Pen-y-clogau, three each at Tair Carn Isaf and Ucaf, Trichug itself and

Figure 99: Ffairfach Grit dipping steeply south-east at the eastern end of Garn Goch

Figure 100: Ffairfach Grit dipping steeply north-west to the west of the barrow.

Day 8. Carreg Cennen to Bethlehem

Figure 101: An original 'postern' entrance to Y Gaer Fawr flanked by upright slabs

Figure 102: Valley formed along the fault between Y Gaer Fawr and Y Gaer Fach. Trichug (Pen y Bicws) is visible behind

on Garn Goch).

After you descend from Pen y Bicws the track northwards follows the fault line described above onto Garn Goch Common. As you cross the Common you should be able to see a low flat topped wooded hill, 2km away to the north-east. This is Coed Duon. The hill consists of volcanic lava and tuff (volcanic ash) erupted from a local volcanic centre in the Welsh Basin around 470 million years ago.

Garn Goch

This site is doubly important, not only is it the largest Prehistoric monument in South Wales, but also the rocks on which and of which it is built are the oldest along the Beacons Way, the 470 million year old Ordovician Ffairfach Grit Formation. These pale grey sandstones, some quite coarse grained, probably originated as debris flows from the shallow margin of the Welsh Basin into deeper water. Such debris flows were often triggered by volcanic eruptions or earthquakes.

The Ffairfach Grits forming the Garn Goch ridge have been folded into a north-east to south-west anticline, providing the perfect site for the two forts, Y Garn Fawr and Y Garn

Figure 103: Ice-smoothed slab of Ffairfach Grit dipping north-west below the rampart of Y Gaer Fawr

109

The Land of the Beacons Way

Fach. Both forts would have been surrounded and defended by tall, massive, coursed stone battlements, now ruined, which followed the contours of the hill. The main path across the common enters Y Garn Fawr by a wide, modern entrance through the north-eastern wall. The original entrances were probably narrow 'postern' gates, eight of which have been found, the best preserved is in the southern wall. Just inside the modern entrance the rocks to your left can be seen dipping steeply to the south-east but to the west of the oval barrow built on the crest of the anticline the beds dip north-west. Due to its rectangular shape this barrow could be Neolithic rather than Bronze Age. Between the two forts there is a shallow valley marking erosion along another strand of the Welsh Borderland Fault System.

From up here there should be a good view north to the Tywi Valley which follows the complex Tywi Lineament, another part of the above fault system. As the path leaves the site it passes the memorial stone to Gwynfor Evans, this is Carboniferous limestone containing many brachiopod fossils, quarried near Llanddybie.

Bethlehem

The walk back through time comes to an end as you leave Garn Goch but this is a spectacular finale. Most of the walk into Bethlehem is over glacial till. The rocks underlying the village are the Llandeilo Flags, still Ordovician in age but slightly younger than the Ffairfach Grits and not exposed along the lane and path.

Figure 104: Y Garn Fawr and the rectangular barrow

Glossary

Acadian Orogeny: the episode of Earth movements and mountain building in the Mid-Devonian Period which created a range of mountains stretching from the NE Appalachians through Western Europe to Poland, and formed the slate belts of England and Wales. In South and Mid Wales its effect was to compress and invert the sediments in the Welsh Basin between 400 and 390 million years ago. It was previously considered to be the last phase of the Caledonian Orogeny but is now thought to be related to collisions in the Rheic Ocean which opened up to the south of Avalonia as the Iapetus ocean closed.

Alluvial fan: a fan shaped sedimentary deposit formed where a river flows off steep terrain onto a flatter plain, slows down and spreads sideways.

Alluvium: loose, unconsolidated sediment which is produced by erosion and then deposited and reshaped by non-marine water.

Anticline: an arched shaped fold where the oldest rocks occur in the core.

Barrier island: an offshore sandbar behind which a lagoon or lake forms.

Bed: the smallest division of a geological formation, distinguished from the layers above and below by bedding surfaces/planes. Beds are usually differentiated by different rock types or grain sizes. Each bed represents a geological event which could have lasted anything from a few minutes to millions of years.

Bioturbation: the process by which the sediment within a bed is mixed up by the burrowing action of animals.

Brachiopods: marine invertebrates with two shells, upper and lower, hinged at the back; also called lamp shells. Most had a stalk protruding from the hinge area which anchored the animal to the sea floor. Common in the Carboniferous period, now mostly extinct but a few species still survive.

Braided river: a river composed of a network of interwoven channels. Such rivers usually have a steep gradient, high water volume, high sediment load and unconsolidated banks. They are also typical of rivers flowing across alluvial fans and deltas.

Break of slope: a change of gradient in the landscape, usually indicating a change in the underlying rock type

Calcretes: nodules or knobbly bands of lime (calcium carbonate, $CaCO_3$), centimetres to metres in thickness. They represent fossil soils and form in semi-arid regions where carbonates dissolve in the seasonal rain, are washed downwards through the loose sediments and then drawn up again by heat and evaporation to be concentrated in lumps near the surface.

Caledonian Orogeny: the episodes of Earth movements and mountain building associated with the closure of the Iapetus Ocean and the subsequent collisions of the continents of Avalonia, Baltica and Laurentia to create the continent of Laurussia. The orogeny consisted of a number of separate events, spanning around 150 million years, starting in the Early Ordovician Period 490 million years ago and ceasing in the mid Devonian Period.

Caprock doline: see doline

Carbonate ramp: a large, gently sloping deposit of calcium carbonate material extending outwards from a shoreline into water still shallow enough for sunlight to penetrate.

Chalk: a fine-grained limestone made from the minute calcium carbonate ($CaCO_3$) skeletal plates (coccoliths) of marine planktonic algae.

Clints: limestone blocks within a limestone pavement bordered by fissures (grikes)

Clubmosses: early plants which appeared in the Devonian Period. By the Carboniferous Period some species had evolved into huge plants up to 40m tall with a simple branching structure and small scale-like leaves attached to the stem by leaf cushions which continued to photosynthesize even after the leaves were shed. They were not true trees as they consisted of soft, non-woody tissue and were anchored by shallow structures called stigmaria which were neither stems nor roots. Stigmaria probably absorbed oxygen directly from the air, an advantage for life in oxygen-poor swamp water. Clubmosses survive today as small creeping plants in nutrient-poor habitats.

Conglomerate: a sedimentary rock made of rounded pebbles of various sizes cemented together.

Continental collision: the collision of two tectonic plates, each carrying continental crust after the oceanic crust flooring the ocean basin between them has been destroyed by subduction under one or both plates. Continental crust is too buoyant to be deeply subducted so the two continental masses ram into each other forcing their rocks upwards to form a mountain chain on greatly thickened continental crust..

Continental drift: the hypothesis describing the slow movement of continents relative to each other over the surface of the Earth (at about the speed at which finger nails grow)

Crinoid: a marine animal, related to starfish and sea urchins, which was usually attached to the sea floor by a stalk and caught minute food with its feeding arms. Some forms still exist but they are rare. They are usually up to 50cm in length but there were some which reached 20m. The stalk consists of ossicles which look like polo mints and are often seen fossilised in limestone rocks. Each ossicle is a single calcium carbonate crystal.

Cross-bedding: laminae within a bed which are inclined relative to the bedding surface. They are created by water or wind flowing over a sandy surface and moulding the sand into ripples or dunes. Sand grains move over the ripple or dune crests to be deposited on the downstream slopes. Each ripple slowly builds out downstream as a set of inclined layers. New ripples build on top of the older ones so that, over time, many sets of inclined layers build up.

Crust: (of the Earth): The outermost, rigid, rocky shell.

Dip: the angle of slope of the bedding surfaces in sedimentary rocks, measured from the horizontal

Dip slope: a topographic surface which slopes in the same direction, and at the same angle, as the dip of the underlying strata. Such surfaces are usually fairly resistant to erosion.

Disturbance: a complex fault zone with a long history of reactivation which has usually moved in both horizontal and vertical directions.

Doline: a depression where water sinks underground.

Solution dolines form on exposed limestone as rain water dissolves the rock and then drains away through fissures at the bottom of the funnel shaped holes. Suffosion dolines form entirely within the overlying loose material, which slowly washes down the underlying fissures. Caprock dolines occur when the overlying insoluble rocks collapse into an underground void in soluble strata. They are the commonest feature where beds of sandstone and grit overlie limestone and are usually called shakeholes in the UK Sink holes form where a stream flowing over insoluble rocks meets fissured limestone and sinks underground and so have water flowing into them, at least at times. In practice most dolines form by a combination of these events and the large examples marked on the OS maps are usually now caprock dolines.

Dolomite: a rock made of the mineral calcium magnesium carbonate.

Erratic: a rock carried by ice and deposited at a distance from its source and on top or amongst of different rocks.

Erosion: the removal of weathered debris which results in lowering of the land surface. Agents of erosion include water, wind, ice and gravity.

Erosion surface: a sharp, often irregular, boundary between two beds produced when the surface of the lower bed is eroded away as the upper bed is deposited.

Fault: a fracture within a mass of rock along which one side has moved relative to the other. In a **normal fault** the mass of rock has been stretched so the rocks on the upper side of the inclined fault plane have dropped downwards. In a **reverse fault** the rocks have been compressed so that older rocks have been pushed over younger ones. A **thrust fault** is a special type of reverse fault where the fault plane is inclined at a low angle.

Fault scarp: a step in the landscape caused by slip on a fault and subsequent erosion to leave the harder strata standing proud. In effect they are fossilised earthquakes.

Flux: a chemical purifying agent added to metal ore during smelting to react with and remove impurities such as silica, carbon, phosphorus and sulphur.

Frost shattering: weathering of rock when water in cracks freezes, expands and breaks the rock apart.

Gastropods: invertebrate animals most of which possess a coiled shell. They can be marine (whelks), fresh-water or terrestrial (snails)

Graben: a structure formed when a block of country is lowered (downthrown), relative to the rocks on each side, between two nearly parallel faults.

Grikes: fissures, widened by rock dissolution, bordering limestone blocks (clints) within a limestone pavement.

Headwall: the headwall of a cirque is the highest of the enclosing rock walls.

Horsetails: early plants with reduced leaves and spores instead of seeds. In the Carboniferous Period they reached 20 – 30 m tall and dominated the forest understorey. Their descendants today are usually less than a metre high.

Hot spot: place at which a huge mass of abnormally hot rock (a mantle plume) rises from deep within the Earth's mantle (possibly even from the core-mantle boundary) towards the surface.

Interstadial: a warm period during a glacial period which is not long enough to be described as an interglacial period. Interstadials last less than 10,000 years.

Intraformational conglomerate: a conglomerate in which the rock fragments were derived from the same geological formation in which they are now found.

Iron ore/Ironstone: in South Wales iron occurs as nodules of iron carbonate within the mud layers of the Coal Measures. These nodules can be small or more than a metre in width. The iron probably came from the decaying vegetation.

Joints: cracks in rocks where there is no movement between the two sides. Joints are usually caused either by earth movements or when the removal of overlying rocks by erosion reduces the pressure on the rock and allows it to expand and crack

Karst: a landscape formed on soluble rocks with an efficient underground drainage. Water sinks underground at dolines so streams and valleys are rare.

Laminae: very thin layers within a bed of sedimentary rock.

Late Devensian glaciation: the last long glacial period to affect the Brecon Beacons area. It started about 26,000 years ago, reached a maximum around 20,000 years ago and started to remit about 15,000 years ago. It followed the Early and Middle Devensian glaciations which lasted from approximately 122, 000 to 26,000 years ago with interspersed brief warmer interstadial episodes.

Limestone: a sedimentary rock made mostly of calcium carbonate ($CaCO_3$) derived either from broken up animal shells or precipitated from water supersaturated in calcium carbonate.

Liquefaction: the process by which a layer of unconsolidated, water saturated sediment looses all its strength and behaves like a liquid. The most common geological situation occurs when earthquake shocks rapidly compress the sediment, increasing the water pressure in the pore spaces between the grains which loose contact with each other causing the whole sediment to become a liquid and flow.

Lithosphere: the outermost layer of the Earth, consisting of all continental and oceanic crust plus the uppermost mantle

Mantle: that part of the Earth's interior between the core and the crust. It consists of solid crystalline rock which can, however, undergo slow, viscous movement at high temperatures and pressures.

Mantle plume: an upwelling of a huge mass of abnormally hot rock from deep within the Earth's mantle.

Mica: a range of silicate minerals in which the molecules are arranged as flat sheets. Micas are usually derived from the weathering of granite.

Moraine: banks of till deposited by ice or ice and water.

Mouth bar: An accumulation of sediment deposited at the mouth of a river channel where fresh and sea water meet where the water flow slows and its chemistry changes.

Mudstone: a sedimentary rock made up of clay sized particles.

Nunatak: an exposed mountain top, not covered by snow or ice, within an ice-field or glacier.

Oolith: round grains, 0.5 – 2 mm in diameter, formed when fragments of shell are rolled around in water saturated with calcium carbonate, which precipitates in concentric layers around the fragments. They resemble masses of tiny eggs in the rock.

Orogeny: the severe deformation and thickening of the Earth's crust to form mountain belts which occurs where tectonic plates converge and collide.

Overdeepened: an area gouged out by ice to a lower level than its outlet.

Palaeokarst: a rock or landscape which was a karst surface in the past, was buried by younger sediments and now exposed again by erosion.

Palaeozoic: the geological era between 542 and 250 million years ago. The name means *early life*.

Patterned Ground: the distinct geometrical patterns formed on the surface in periglacial environments. Their cause is complex but needs either permafrost conditions or alternate freezing and thawing. Freezing of wet, fine grained material increases its volume by 10% so the ground heaves upwards. This frost heave sorts the stones by size with the larger ones moving upwards as finer material flows downwards between them. Thawing results in theses larger stones on the raised ground sliding off to form patterns (polygons, circles, nets or stripes), depending on the stone size, slope of the surface and the frequency and duration of the freeze-thaw cycles..

Periglacial: the landforms and climate of an area experiencing long-standing cold but non-glacial conditions and where freeze-thaw weathering of rock occurs.

Pronivial rampart: a mound or ridge formed by debris sliding down a long lived snowbed and accumulating at the foot.

Pyroclastic flow: a searingly hot, turbulent mixture of superheated gases, ash and rocks, usually produced by the gravitational collapse of a volcanic eruption column. These flows travel down the volcano's sides at > 100mph destroying everything in their path.

Reduction spot: a greenish area within Old Red Sandstone rocks where the red ferric iron has been reduced to green ferrous iron, usually by the oxidative decay of plant or animal material.

Ripple marks: wavy ridges formed by currents or waves and preserved on a sediment surface.

Rottenstone: colloquial name for the youngest layer of the Carboniferous limestone when it has been invaded by silicaceous fluids from the overlying Twrch Sandstone, and later weathered. The fluids crystallized as a framework of hard silica around the limestone grains. The latter then dissolved by weathering leaving a friable rock which is easily crushed to make a coarse powder suitable for polishing metal, on both industrial and domestic scales.

Sandstone: a sedimentary rock composed mainly of cemented sand grains

Sedimentary rocks: rocks formed from material deposited on the Earth's surface.

Shakehole: see **doline.**

Shale: a mudstone in which the clay particles are laid down in flat layers, so that the rock splits into thin layers

Silica: silicon dioxide, SiO_2. Minerals made from silica form over 90% of the Earth's crust. Sand and sandstone are primarily silica.

Silica bricks: bricks made of about 95% silica plus small amounts of aluminium and lime. They were produced to withstand the very high temperatures in blast furnaces.

Silica sand: sand which is made up of > 95% silicon dioxide, therefore with low levels of impurities. The silica sand referred to here was formed by the disaggregation (physical and chemical weathering) of silica rock, probably under tropical conditions during the Palaeogene and Neogene Periods (66-2.6 million years ago). It was used for manufacturing refractory furnace bricks, foundry castings and glass.

Silt: a sedimentary rock composed of particles intermediate in size between clay and sand.

Sink hole: see doline

Solifluction: the slow downslope movement of water-saturated material in periglacial conditions.

Solution doline: see doline.

Stigmaria: the anchoring system of the giant clubmoss *Lepidodendron*. They are neither stems nor roots but a type of rhizome from which grew masses of small, straight rootlets. The impressions left by these rootlets are fossilized in sediments.

Strike: the orientation, clockwise from north, of a bedding plane

Subduction: the sinking of one tectonic plate below another where two plates meet.

Suffusion doline: see doline

Supercontinent: a huge single continent formed of all, or nearly all, the Earth's landmasses.

Suture line: the zone where two tectonic plates have collided and welded together.

Syncline: a downfold in bedded rocks, producing a trough-like structure.

Tectonic plates: huge pieces of the Earth's crust and uppermost mantle (the lithosphere) which ride over the Earth's surface on a weak layer in the mantle (the asthenosphere). Their driving force is thought to be the sinking into the mantle of old, cold, dense lithosphere; a process called slab pull. At present there are seven major and eight minor plates, each of which moves relative to its neighbours.

Till: material deposited by glaciers, ranging in size from clay to boulders.

Thrust: a low-angle reverse fault (see fault)

Trace fossil: impressions left in sediments by the activities of living organisms

Trimline: a boundary line on the side of a glacially formed valley marking the upper surface of the last ice sheet to occupy that valley at its maximum thickness.

Unconformity: a time gap in the rock sequence indicating non-deposition or erosion

Variscan Orogeny: the episode of earth movements and mountain building lasting from the Late Devonian, through the Carboniferous and into the Early Permian Period, during which the continents of Laurussia and Gondwana collided to form the supercontinent of Pangea.

Vascular plant: a plant with specialised channels for transporting water, minerals and sugars around the plant. They first appeared in the Silurian Period but did not diversify and become large plants until the Devonian.

Weathering: the break-up of rocks on the Earth's surface by physical or chemical agents.

Younger Dryas Stadial: the short lived cold episode between about 12,500 and 11,500 years ago which produced small glaciers and perennial snowbeds beneath some escarpments in the Brecon Beacons area.

Further Information

About British geology

Brenchley, P.J. & Rawson, P.F., 2006. *The Geology of England and Wales*. London: The Geological Society.

Toghill, P., 2002. *The Geology of Britain*. Airlife Publishing.

Woodcock, N. & Strachan, R. (eds), 2012. *Geological History of Britain and Ireland*, 2nd edn. Wiley-Blackwell.

About the geology and archaeology of South Wales

Barclay, W.J. & Wilby, P.R., 2003. *Geology of the Talgarth district*. British Geological Survey

Barclay, W J, Davies, J R, Humpage, A J, Waters, R A, Wilby, P R, Williams, M and Wilson, D. 2005. *Geology of the Brecon district - a brief explanation of the geological map Sheet 213 Brecon. Sheet explanation of the British Geological Survey. 1:50000 Sheet 213 Brecon*

British Geological Survey, 2014. Fforest Fawr Exploring the Landscape of a Global Geopark. 1:50,000 scale map with much geological information and the western section of the Beacons Way marked

Burgess, P., 2010. *Penwyllt, The story of a South Wales Community*. Published by the author.

Burrow, S., 2011. *Shadowland, Wales 3000-1500 BC*. Amgueddfa Cymru-National Museum of Wales & Oxbow Books

Carr, S.J., Coleman, C.G., Humpage, A.J., Shakesby., R.A. 2007. *Quaternary of the Brecon Beacons Field Guide*. Quaternary Research Association

George, G.T., 2008. *The Geology of South Wales: A Field Guide*. Published by the author, gareth@geoserv.co.uk

Howells, M.F., 2007. *British Regional Geology Wales*. British Geological Survey

Hughes, S., 1990. *The Brecon Forest Tramroads*. RCAHMW

Leighton, D., 2012. *The Western Brecon Beacons. The Archaeology of Mynydd Du and Fforest Fawr*. RCAHMW

Owen, T.R., 1973. *Geology explained in South Wales*. Fineleaf Editions

Schofield, D.I.; Davies, J.R.; Jones, N.S.; Leslie, A.B.; Waters, R.A.; Williams, M.; Wilson, D.; Venus, J.; Hillier, R.D.. 2009 Geology of the Llandovery district : a brief explanation of the geological map Sheet 212 Llandovery. Nottingham, UK, British Geological Survey, 38pp. (Explanation (England and Wales Sheet) British Geological Survey, 212)

Shakesby, R., 2002. *Classic Landforms of the Brecon Beacons*. Geographical Association

Thomas, B.A. & Cleal, C.J., 2000. *Invasion of the Land*. National Museums & Galleries of Wales. Amgueddfeydd ac Orielau Cenedlaethol Cymru

Thomas, B.A. & Cleal, C.J., 1993. *The Coal Measures Forests*. National Museum of Wales. Amgueddfa Genedlaethol Cymru

Fforest Fawr Geopark Geotrails, available from Information Centres

Garn Goch

Glyn Tarell

British Geological Survey apps (free)

iGeology

iGeology3D

Acknowledgements

This would not have been written without the encouragement and help of Alan Bowring (Fforest Fawr Geopark Development Officer), Rhian Kendall (editor South Wales Group of the Geologists' Association) and Geraint Owen (Department of Geography University of Swansea).

I would also like to thank Louise Austin (Dyfed Archaeological Trust), John Davies (Welsh Stone Forum), Lynda Garfield (South Wales Group of the Geologists' Association), Dave Green (geology tutor), Jonathan Harlow, Duncan Hawley (Swansea Metropolitan University), Stephen Howe (South Wales Group of the Geologists' Association), Adrian Humpage (South East Wales RIGS Group), Andy Kendall, Ruth Palmer (Brecon Beacons Park Society), Tony Ramsay (Scientific Director Fforest Fawr Geopark), Duncan Schlee (Dyfed Archaeological Trust), David Schofield (British Geological Survey Cardiff) and Soo Turnbull (Brecon Beacons Park Society) for their helpful comments, information and advice.

Figures

All of the images are by the author unless listed below:

Fig 1, 2, 3 and 23 Scotese, C.R., 2001. Atlas of Earth History, Volume 1, Paleogeography, PALEOMAP Project, Arlington, Texas, 25 pp

Fig 4 by permission of the British Geological Survey © NERC. All rights reserved

Fig 6 Rhian Kendall after Edwards, D. 1970. Fertile rhyniophytina from the Lower Devonian of Britain. Palaeontology 13(3), pp. 451-461

Fig 9 E M Bridges. World Soils 1997. Reproduced with permission from Cambridge University Press

Fig 10 and 16b courtesy of Wikimedia Commons

Fig 11 and 27 Alan Bowring

Fig 12, 15, 20, 28, 68 and 73 Rhian and Andy Kendall

Fig 16a Rhian kendall after South Wales Geologists Association Walking the Rocks

Fig 21 Geoffrey Williams, Brecon Beacons Park Society

Fig 22 Annette Townsend, Department of Biodiversity and Systematic Biology Amguedda Cymru, from Invasion of the Land published by Amuuedda Cymru; reproduced with permission

Fig 24, 25, 29 and 95 Rhian Kendall after T R Owen Geology Explained in South Wales

Fig 26 Jez Everest, British Geological Survey

Fig 29 Rhian Kendall after Tony Waltham, 1987. Karst and Caves - Yorkshire Dales National Park / Topic Series

Fig 34 Robert Reith, National Trust

Fig 50 Rhian Kendall after Dineley, D & Metcalf, S., 1999. Fossil Fishes of Great Britain, Geological Conservation Review Series

The Land of the Beacons Way

Notes

The Land of the Beacons Way

Notes

The Land of the Beacons Way

Notes

The Land of the Beacons Way

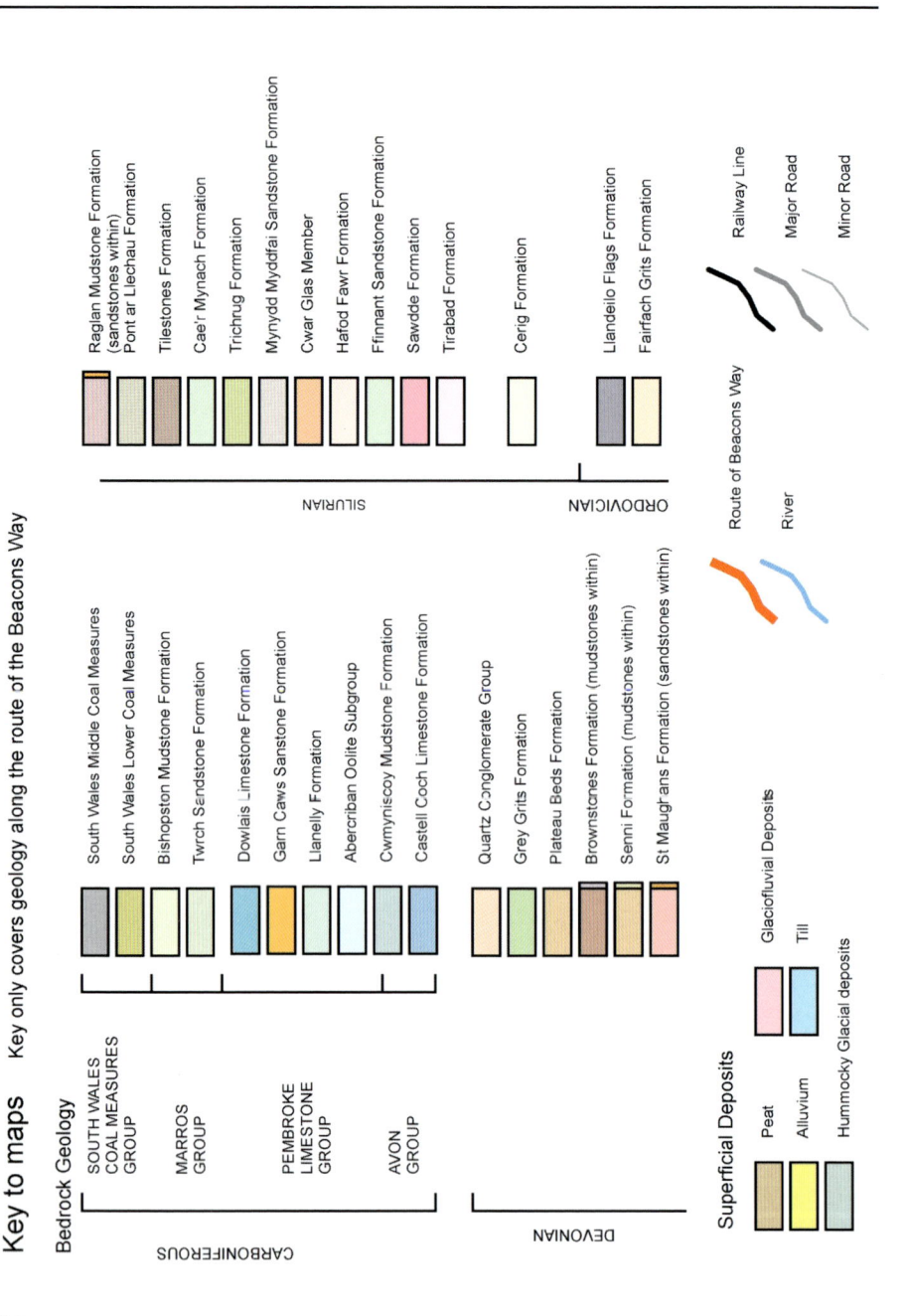

Key to maps — Key only covers geology along the route of the Beacons Way

Bedrock Geology

CARBONIFEROUS

SOUTH WALES COAL MEASURES GROUP
- South Wales Middle Coal Measures
- South Wales Lower Coal Measures

MARROS GROUP
- Bishopston Mudstone Formation
- Twrch Sandstone Formation

PEMBROKE LIMESTONE GROUP
- Dowlais Limestone Formation
- Garn Caws Sanstone Formation
- Llanelly Formation
- Abercriban Oolite Subgroup

AVON GROUP
- Cwmyniscoy Mudstone Formation
- Castell Coch Limestone Formation

DEVONIAN
- Quartz Conglomerate Group
- Grey Grits Formation
- Plateau Beds Formation
- Brownstones Formation (mudstones within)
- Senni Formation (mudstones within)
- St Maughans Formation (sandstones within)

SILURIAN
- Raglan Mudstone Formation (sandstones within)
- Pont ar Llechau Formation
- Tilestones Formation
- Cae'r Mynach Formation
- Trichrug Formation
- Mynydd Myddfai Sandstone Formation
- Cwar Glas Member
- Hafod Fawr Formation
- Ffinnant Sandstone Formation
- Sawdde Formation
- Tirabad Formation

ORDOVICIAN
- Cerig Formation
- Llandeilo Flags Formation
- Fairfach Grits Formation

Superficial Deposits
- Peat
- Alluvium
- Hummocky Glacial deposits
- Glaciofluvial Deposits
- Till

Route of Beacons Way
River
Railway Line
Major Road
Minor Road

128